SHORE PROTECTION MANUAL

VOLUME III

(Appendixes A Through D)

U.S. ARMY COASTAL ENGINEERING RESEARCH CENTER

Books for Business
New York-Hong Kong

Shore Protection Manual
(Volume Three)

by
U. S. Army Coastal Engineering Research
Center

ISBN: 0-89499-099-3

Reprinted from the 1973 edition

Books for Business
New York - Hong Kong
http://www.BusinessBooksInternational.com

TABLE OF CONTENTS

VOLUME I

VOLUME II

TABLE OF CONTENTS

VOLUME III

APPENDIX

A P P E N D I X   A

GLOSSARY OF TERMS

The glossary was compiled and reviewed by the Staff of the Coastal Engineering Research Center. The terms came from many sources. However, the following publications were of particular value.

American Geological Institute (1957) - *Glossary of Geology and Related Sciences with Supplement*. 2d Edition.

American Meteorological Society (1959) - *Glossary of Meteorology*.

U.S. Army Coastal Engineering Research Center (1966) - *Shore Protection, Planning and Design*, Technical Report No. 4, 3d Edition.

U.S. Coast and Geodetic Survey (1949) - *Tide and Current Glossary*, Special Publication No. 228, Revised (1949) Edition.

U.S. Navy Oceanographic Office (1966) - *Glossary of Oceanographic Terms*, Special Publication (SP-35), 2d Edition.

Wiegel, R.L. (1953) - *Waves, Tides, Currents and Beaches: Glossary of Terms and List of Standard Symbols*. Council on Wave Research, The Engineering Foundation, University of California.

NEWPORT COVE, MAINE -- 12 September, 1958

APPENDIX A

GLOSSARY

OF

TERMS

APPENDIX A

GLOSSARY OF TERMS

ACCRETION - May be either NATURAL or ARTIFICIAL. Natural accretion is the buildup of land, solely by the action of the forces of nature, on a BEACH by deposition of waterborne or airborne material. Artificial accretion is a similar buildup of land by reason of an act of man, such as the accretion formed by a groin, breakwater, or beach fill deposited by mechanical means. Also AGGRADATION.

ADVANCE (OF A BEACH) - (1) A continuing seaward movement of the shoreline. (2) A net seaward movement of the shoreline over a specified time. Also PROGRESSION.

AGE, WAVE - The ratio of wave velocity to wind velocity (in wave forecasting theory).

AGGRADATION - See ACCRETION.

ALLUVIUM - Soil (sand, mud, or similar detrital material) deposited by streams, or the deposits formed.

ALONGSHORE - Parallel to and near the shoreline; same as LONGSHORE.

AMPLITUDE, WAVE - (1) The magnitude of the displacement of a wave from a mean value. An ocean wave has an amplitude equal to the vertical distance from stillwater level to wave crest. For a sinusoidal wave, amplitude is one-half the wave height. (2) The semirange of a constituent tide.

ANTIDUNES - BED FORMS that occur in trains, and are in phase with and strongly interact with gravity water-surface waves.

ANTINODE - See LOOP.

ARTIFICIAL NOURISHMENT - The process of replenishing a beach with material (usually sand) obtained from another location.

ATOLL - A ring-shaped coral reef, often carrying low sand islands, enclosing a lagoon.

ATTENUATION - (1) A lessening of the amplitude of a wave with distance from the origin. (2) The decrease of water-particle motion with increasing depth. Particle motion resulting from surface oscillatory waves attenuates rapidly with depth, and practically disappears at a depth equal to a surface wavelength.

AWASH - Situated so that the top is intermittently washed by waves or tidal action. Condition of being exposed or just bare at any stage of the tide between high water and chart datum.

BACKBEACH - See BACKSHORE.

BACKRUSH - The seaward return of the water following the uprush of the
waves. For any given tide stage the point of farthest return seaward of the backrush is known as the LIMIT of BACKRUSH or LIMIT
BACKWASH. (See Figure A-2.)

BACKSHORE - That zone of the shore or beach lying between the foreshore
and the coastline and acted upon by waves only during severe storms,
especially when combined with exceptionally high water. Also
BACKBEACH. It comprises the BERM or BERMS. (See Figure A-1.)

BACKWASH - (1) See BACKRUSH. (2) Water or waves thrown back by an
obstruction such as a ship, breakwater, or cliff.

BANK - (1) The rising ground bordering a lake, river, or sea; of a
river or channel, designated as right or left as it would appear
facing downstream. (2) An elevation of the sea floor of large
area, located on a Continental (or island) Shelf and over which
the depth is relatively shallow but sufficient for safe surface
navigation; a group of shoals. (3) In its secondary sense, a
shallow area consisting of shifting forms of silt, sand, mud, and
gravel, but in this case it is only used with a qualifying word
such as "sandbank" or "gravelbank".

BAR - A submerged or emerged embankment of sand, gravel, or other unconsolidated material built on the sea floor in shallow water by
waves and currents. See BAYMOUTH BAR, CUSPATE BAR. (See Figures
A-2 and A-9.)

BARRIER BEACH - A bar essentially parallel to the shore, the crest of
which is above normal high water level. Also called OFFSHORE
BARRIER and BARRIER ISLAND. (See Figure A-9.)

BARRIER LAGOON - A bay roughly parallel to the coast and separated from
the open ocean by barrier islands. Also the body of water encircled
by coral islands and reefs, in which case it may be called an atoll
lagoon.

BARRIER REEF - A coral reef parallel to and separated from the coast by
a lagoon that is too deep for coral growth. Generally, barrier
reefs follow the coasts for long distances, and are cut through at
irregular intervals by channels or passes.

BASIN, BOAT - A naturally or artificially enclosed or nearly enclosed
harbor area for small craft.

BATHYMETRY - The measurement of depths of water in oceans, seas, and
lakes; also information derived from such measurements.

BAY - A recess in the shore or an inlet of a sea between two capes or headlands, not as large as a gulf but larger than a cove. See also BIGHT, EMBAYMENT. (See Figure A-9.)

BAYMOUTH BAR - A bar extending partly or entirely across the mouth of a bay. (See Figure A-9.)

BAYOU - A minor sluggish waterway or estuarial creek, tributary to, or connecting, other streams or bodies of water. Its course is usually through lowlands or swamps. Sometimes called SLOUGH.

BEACH - The zone of unconsolidated material that extends landward from the low water line to the place where there is marked change in material or physiographic form, or to the line of permanent vegetation (usually the effective limit of storm waves). The seaward limit of a beach - unless otherwise specified - is the mean low water line. A beach includes FORESHORE and BACKSHORE. (See Figure A-1.)

BEACH ACCRETION - See ACCRETION.

BEACH BERM - A nearly horizontal part of the beach or backshore formed by the deposit of material by wave action. Some beaches have no berms, others have one or several. (See Figure A-1.)

BEACH CUSP - See CUSP.

BEACH EROSION - The carrying away of beach materials by wave action, tidal currents, littoral currents, or wind.

BEACH FACE - The section of the beach normally exposed to the action of the wave uprush. The FORESHORE of a BEACH. (Not synonymous with SHOREFACE.) (See Figure A-2.)

BEACH RIDGE - See RIDGE, BEACH.

BEACH SCARP - See SCARP, BEACH.

BEACH WIDTH - The horizontal dimension of the beach measured normal to the shoreline.

BED FORMS - Any deviation from a flat bed that is readily detectable by eye, and higher than the largest sediment size present in the parent bed material; generated on the bed of an alluvial channel by the flow.

BEDLOAD - See LOAD.

BENCH - (1) A level or gently sloping erosion plane inclined seaward. (2) A nearly horizontal area at about the level of maximum high water on the sea side of a dike.

BENCH MARK - A permanently fixed point of known elevation. A primary
    bench mark is one close to a tide station to which the tide staff
    and tidal datum originally are referenced.

BERM, BEACH - See BEACH BERM.

BERM CREST - The seaward limit of a berm. Also BERM EDGE. (See
    Figure A-1.)

BIGHT - A bend in a coastline forming an open bay. A bay formed by
    such a bend. (See Figure A-8.)

BLOWN SANDS - See EOLIAN SANDS.

BLUFF - A high steep bank or cliff.

BOLD COAST - A prominent land mass that rises steeply from the sea.

BORE - A very rapid rise of the tide in which the advancing water pre-
    sents an abrupt front of considerable height. In shallow estuaries
    where the range of tide is large, the high water is propagated in-
    ward faster than the low water because of the greater depth at high
    water. If the high water overtakes the low water, an abrupt front
    is presented with the high water crest finally falling forward as
    the tide continues to advance. Also EAGER.

BOTTOM - The ground or bed under any body of water; the bottom of the
    sea. (See Figure A-1.)

BOTTOM (NATURE OF) - The composition or character of the bed of an ocean
    or other body of water (e.g., clay, coral, gravel, mud, ooze,
    pebbles, rock, shell, shingle, hard, or soft).

BOULDER - A rounded rock more than 10 inches in diameter; larger than a
    cobblestone. See SOIL CLASSIFICATION.

BREAKER - A wave breaking on a shore, over a reef, etc. Breakers may be
    classified into four types (see Figure A-4):

    Spilling - bubbles and turbulent water spill down front face of
        wave. The upper 25 percent of the front face may become
        vertical before breaking. Breaking generally across over
        quite a distance.

    Plunging - crest curls over air pocket; breaking is usually with
        a crash. Smooth splash-up usually follows.

    Collapsing - breaking occurs over lower half of wave. Minimal air
        pocket and usually no splash-up. Bubbles and foam present.
        (See Figure 2-69.)

Surging - wave peaks up, but bottom rushes forward from under wave, and wave slides up beach face with little or no bubble production. Water surface remains almost plane except where ripples may be produced on the beachface during runback.

BREAKER DEPTH - The stillwater depth at the point where a wave breaks. Also BREAKING DEPTH. (See Figure A-2.)

BREAKWATER - A structure protecting a shore area, harbor, anchorage, or basin from waves.

BULKHEAD - A structure or partition to retain or prevent sliding of the land. A secondary purpose is to protect the upland against damage from wave action.

BUOY - A float; especially a floating object moored to the bottom, to mark a channel, anchor, shoal, rock, etc.

BUOYANCY - The resultant of upward forces, exerted by the water on a submerged or floating body, equal to the weight of the water displaced by this body.

BYPASSING, SAND - Hydraulic or mechanical movement of sand from the accreting updrift side to the eroding downdrift side of an inlet or harbor entrance. The hydraulic movement may include natural as well as movement caused by man.

CANAL - An artificial watercourse cut through a land area for such uses as navigation and irrigation.

CANYON - A relatively narrow, deep depression with steep slopes, the bottom of which grades continuously downward. May be underwater (submarine) or on land (subaerial).

CAPE - A relatively extensive land area jutting seaward from a continent or large island which prominently marks a change in, or interrupts notably, the coastal trend; a prominent feature. (See Figure A-8.)

CAPILLARY WAVE - A wave whose velocity of propagation is controlled primarily by the surface tension of the liquid in which the wave is traveling. Water waves of length less than about 1 inch are considered capillary waves. Waves longer than 1 inch and shorter than 2 inches are in an indeterminate zone between CAPILLARY and GRAVITY WAVES. See RIPPLE.

CAUSEWAY - A raised road, across wet or marshy ground, or across water.

CAUSTIC - In refraction of waves, the name given to the curve to which adjacent orthogonals of waves refracted by a bottom whose contour lines are curved, are tangents. The occurrence of a caustic always marks a region of crossed orthogonals and high wave convergence.

CAY - See KEY.

CELERITY - Wave speed.

CENTRAL PRESSURE INDEX (CPI) - The estimated minimum barometric pressure
in the *eye (approximate center)* of a particular hurricane. The CPI
is considered the most stable index to intensity of hurricane wind
velocities in the periphery of the storm; the highest wind speeds
are associated with storms having the lowest CPI.

CHANNEL - (1) A natural or artificial waterway of perceptible extent
which either periodically or continuously contains moving water,
or which forms a connecting link between two bodies of water.
(2) The part of a body of water deep enough to be used for
navigation through an area otherwise too shallow for navigation.
(3) A large strait, as the English Channel. (4) The deepest
part of a stream, bay, or strait through which the main volume
or current of water flows.

CHARACTERISTIC WAVE HEIGHT - See SIGNIFICANT WAVE HEIGHT.

CHART DATUM - The plane or level to which soundings (or elevations) or
tide heights are referenced (usually LOW WATER DATUM). The surface
is called a tidal datum when referred to a certain phase of tide.
To provide a safety factor for navigation, some level lower than
MEAN SEA LEVEL is generally selected for hydrographic charts such
as MEAN LOW WATER or MEAN LOWER LOW WATER. See DATUM PLANE.

CHOP - The short-crested waves that may spring up quickly in a moderate
breeze, and break easily at the crest. Also WIND CHOP.

CLAPOTIS - The French equivalent for a type of STANDING WAVE. In American
usage it is usually associated with the standing wave phenomenon
caused by the reflection of a nonbreaking wave train from a structure
with a face that is vertical or nearly vertical. Full clapotis is one
with 100 percent reflection of the incident wave; partial clapotis is
one with less than 100 percent reflection.

CLAY - See SOIL CLASSIFICATION.

CLIFF - A high, steep face of rock; a precipice. See also SEA CLIFF.

CNOIDAL WAVE - A type of wave in shallow water (depth of water is less
than 1/8 to 1/10 the wavelength). The surface profile is expressed
in terms of the Jacobian elliptic function *cn u*; hence the term
cnoidal.

COAST - A strip of land of indefinite width (may be several miles) that
extends from the shoreline inland to the first major change in
terrain features. (See Figure A-1.)

A-6

COASTAL AREA - The land and sea area bordering the shoreline. (See Figure A-1.)

COASTAL PLAIN - The plain composed of horizontal or gently sloping strata of clastic materials fronting the coast, and generally representing a strip of sea bottom that has emerged from the sea in recent geologic time.

COASTLINE - (1) Technically, the line that forms the boundary between the COAST and the SHORE. (2) Commonly, the line that forms the boundary between the land and the water.

COBBLE (COBBLESTONE) - See SOIL CLASSIFICATION.

COMBER - (1) A deepwater wave whose crest is pushed forward by a strong wind; much larger than a whitecap. (2) A long-period breaker.

CONTINENTAL SHELF - The zone bordering a continent and extending from the low water line to the depth (usually about 100 fathoms) where there is a marked or rather steep descent toward a greater depth.

CONTOUR - A line on a map or chart representing points of equal elevation with relation to a DATUM. It is called an ISOBATH when connecting points of equal depth below a datum.

CONTROLLING DEPTH - The least depth in the navigable parts of a waterway, governing the maximum draft of vessels that can enter.

CONVERGENCE - (1) In refraction phenomena, the decreasing of the distance between orthogonals in the direction of wave travel. Denotes an area of increasing wave height and energy concentration. (2) In wind-setup phenomena, the increase in setup observed over that which would occur in an equivalent rectangular basin of uniform depth, caused by changes in planform or depth; also the decrease in basin width or depth causing such increase in setup.

CORAL - (1) (Biology) Marine coelenterates (Madreporaria), solitary or colonial, which form a hard external covering of calcium compounds, or other materials. The corals which form large reefs are limited to warm, shallow waters, while those forming solitary, minute growths may be found in colder waters to great depths. (2) (Geology) The concretion of coral polyps, composed almost wholly of calcium carbonate, forming reefs, and tree-like and globular masses. May also include calcareous algae and other organisms producing calcareous secretions, such as bryozoans and hydrozoans.

CORE - A vertical cylindrical sample of the bottom sediments from which the nature and stratification of the bottom may be determined.

COVE - A small, sheltered recess in a coast, often inside a larger embayment. (See Figure A-8.)

CREST LENGTH, WAVE - The length of a wave *along* its crest. Sometimes called CREST WIDTH.

CREST OF BERM - The seaward limit of a berm. Also BERM EDGE. (See Figure A-1.)

CREST OF WAVE - (1) the highest part of a wave. (2) That part of the wave above stillwater level. (See Figure A-3.)

CREST WIDTH, WAVE - See CREST LENGTH, WAVE.

CURRENT - A flow of water.

CURRENT, COASTAL - One of the offshore currents flowing generally parallel to the shoreline in the deeper water beyond and near the surf zone. They are not related genetically to waves and resulting surf, but may be related to tides, winds, or distribution of mass.

CURRENT, DRIFT - A broad, shallow, slow-moving ocean or lake current. Opposite of CURRENT, STREAM.

CURRENT, EBB - The tidal current away from shore or down a tidal stream. Usually associated with the decrease in the height of the tide.

CURRENT, EDDY - See EDDY.

CURRENTS, FEEDER - The parts of the NEARSHORE CURRENT SYSTEM that flow parallel to shore before converging and forming the neck of the RIP CURRENT.

CURRENT, FLOOD - The tidal current toward shore or up a tidal stream. Usually associated with the increase in the height of the tide.

CURRENT, INSHORE - See INSHORE CURRENT.

CURRENT, LITTORAL - Any current in the littoral zone caused primarily by wave action, e.g., longshore current, rip current. See also CURRENT, NEARSHORE.

CURRENT, LONGSHORE - The littoral current in the breaker zone moving essentially parallel to the shore, usually generated by waves breaking at an angle to the shoreline.

CURRENT, NEARSHORE - A current in the NEARSHORE ZONE. See Figure A-1.

CURRENT, OFFSHORE - See OFFSHORE CURRENT.

CURRENT, PERIODIC - See CURRENT, TIDAL.

CURRENT, PERMANENT - See PERMANENT CURRENT.

CURRENT, RIP - See RIP CURRENT.

CURRENT, STREAM - A narrow, deep, and swift ocean current, as the Gulf Stream. Opposite of CURRENT, DRIFT.

CURRENT SYSTEM, NEARSHORE - See NEARSHORE CURRENT SYSTEM.

CURRENT, TIDAL - The alternating horizontal movement of water associated with the rise and fall of the tide caused by the astronomical tide-producing forces. Also CURRENT, PERIODIC. See also CURRENT, FLOOD and CURRENT, EBB.

CUSP - One of a series of low mounds of beach material separated by crescent-shaped troughs spaced at more or less regular intervals along the beach face. Also BEACH CUSP. (See Figure A-7.)

CUSPATE BAR - A crescent-shaped bar uniting with the shore at each end. It may be formed by a single spit growing from shore and then turning back to again meet the shore, or by two spits growing from the shore and uniting to form a bar of sharply cuspate form. (See Figure A-9.)

CYCLOIDAL WAVE - A steep, symmetrical wave whose crest forms an angle of 120 degrees. The wave form is that of a cycloid. A trochoidal wave of maximum steepness. See also TROCHOIDAL WAVE.

DAILY RETARDATION (OF TIDES) - The amount of time by which corresponding tidal phases grow later day by day (about 50 minutes).

DATUM, CHART - See CHART DATUM.

DATUM, PLANE- The horizontal plane to which soundin s, ground elevations. or water surface elevations are referred. Also REFERENCE PLANE. The plane is called a TIDAL DATUM when defined by a certain phase of the tide. The following datums are ordinarily used on hydrographic charts:

> MEAN LOW WATER - Atlantic coast (U. S.), Argentina, Sweden,
>   and Norway;
> MEAN LOWER LOW WATER - Pacific coast (U. S.);
> MEAN LOW WATER SPRINGS - United Kingdom, Germany, Italy, Brazil,
>   and Chile;
> LOW WATER DATUM - Great Lakes (U. S. and Canada):
> LOWEST LOW WATER SPRINGS - Portugal;
> LOW WATER INDIAN SPRINGS - India and Japan (See INDIAN TIDE
>   PLANE):
> LOWEST LOW WATER - France, Spain, and Greece.

A common datum used on topographic maps is based on MEAN SEA LEVEL. See also BENCH MARK.

DEBRIS LINE - A line near the limit of storm wave uprush marking the landward limit of debris deposits.

DECAY DISTANCE - The distance waves travel after leaving the generating area (FETCH).

DECAY OF WAVES - The change waves undergo after they leave a generating area (FETCH) and pass through a calm, or region of lighter winds. In the process of decay, the significant wave height decreases and the significant wavelength increases.

DEEP WATER - Water so deep that surface waves are little affected by the ocean bottom. Generally, water deeper than one-half the surface wavelength is considered deep water.

DEFLATION - The removal of loose material from a beach or other land surface by wind action.

DELTA - An alluvial deposit, roughly triangular or digitate in shape, formed at a river mouth.

DEPTH - The vertical distance from a specified tidal datum to the sea floor.

DEPTH OF BREAKING - The stillwater depth at the point where the wave breaks. Also BREAKER DEPTH (See Figure A-2.)

DEPTH CONTOUR - See CONTOUR.

DEPTH, CONTROLLING - See CONTROLLING DEPTH.

DEPTH FACTOR - See SHOALING COEFFICIENT.

DERRICK STONE - See STONE, DERRICK.

DESIGN HURRICANE - See HYPOTHETICAL HURRICANE.

DIFFRACTION (of water waves) - The phenomenon by which energy is transmitted laterally along a wave crest. When a part of a train of waves is interrupted by a barrier, such as a breakwater, the effect of diffraction is manifested by propagation of waves into the sheltered region within the barrier's geometric shadow.

DIKE(DYKE) - A wall or mound built around a low-lying area to prevent flooding.

DIURNAL - Having a period or cycle of approximately one TIDAL DAY.

DIURNAL TIDE - A tide with one high water and one low water in a tidal day. (See Figure A-10.)

DIVERGENCE - (1) In refraction phenomena, the increasing of distance between orthogonals in the direction of wave travel. Denotes an area of decreasing wave height and energy concentration. (2) In wind-setup phenomena, the decrease in setup observed under that which would occur in an equivalent rectangular basin of uniform depth, caused by changes in planform or depth. Also the increase in basin width or depth causing such decrease in setup.

DOLPHIN - A cluster of piles.

DOWNCOAST - In United States usage, the coastal direction generally trending toward the south.

DOWNDRIFT - The direction of predominant movement of littoral materials.

DRIFT (noun) - (1) Sometimes used as a short form for LITTORAL DRIFT (2) The speed at which a current runs. (3) Also floating material deposited on a beach (driftwood). (4) A deposit of a continental ice sheet, as a drumlin.

DRIFT CURRENT - A broad, shallow, slow-moving ocean or lake current.

DUNES - (1) Ridges or mounds of loose, wind-blown material, usually sand. (See Figure A-7.) (2) BED FORMS smaller than bars but larger than ripples that are out of phase with any water-surface gravity waves associated with them.

DURATION - In wave forecasting, the length of time the wind blows in nearly the same direction over the FETCH (generating area).

DURATION, MINIMUM - The time necessary for steady-state wave conditions to develop for a given wind velocity over a given fetch length.

EAGER - See BORE.

EBB CURRENT - The tidal current away from shore or down a tidal stream; usually associated with the decrease in the height of the tide.

EBB TIDE - The period of tide between high water and the succeeding low water; a falling tide. (See Figure A-10.)

ECHO SOUNDER - An electronic instrument used to determine the depth of water by measuring the time interval between emission of a sonic or ultrasonic signal and the return of its echo from the bottom.

EDDY - A circular movement of water formed on the side of a main current. Eddies may be created at points where the main stream passes projecting obstructions or where two adjacent currents flow counter to each other.

EDDY CURRENT - See EDDY.

EDGE WAVE - An ocean wave parallel to a coast, with crests normal to the
    shoreline. An edge wave may be standing or progressive. Its height
    diminishes rapidly seaward and is negligible at a distance of one
    wavelength offshore.

EMBANKMENT - An artificial bank such as a mound or dike, generally built
    to hold back water or to carry a roadway.

EMBAYED - Formed into a bay or bays, as an embayed shore.

EMBAYMENT - An indentation in the shoreline forming an open bay.

ENERGY COEFFICIENT - The ratio of the energy in a wave per unit crest
    length transmitted forward with the wave at a point in shallow water
    to the energy in a wave per unit crest length transmitted forward
    with the wave in deep water. On refraction diagrams this is equal
    to the ratio of the distance between a pair of orthogonals at a
    selected point to the distance between the same pair of orthogonals
    in deep water. Also the square of the REFRACTION COEFFICIENT.

ENTRANCE - The avenue of access or opening to a navigable channel.

EOLIAN SANDS - (or BLOWN SANDS) - Sediments of sand size or smaller which
    have been transported by winds. They may be recognized in marine
    deposits off desert coasts by the greater angularity of the grains
    compared with waterborne particles.

EROSION - The wearing away of land by the action of natural forces. On
    a beach, the carrying away of beach material by wave action, tidal
    currents, littoral currents, or by deflation.

ESCARPMENT - A more or less continuous line of cliffs or steep slopes
    facing in one general direction which are caused by erosion or
    faulting. Also SCARP. (See Figure A-1.)

ESTUARY - (1) The part of a river that is affected by tides. (2) The
    region near a river mouth in which the fresh water of the river
    mixes with the salt water of the sea.

EYE - In meteorology, usually the "eye of the storm" (hurricane); the
    roughly circular area of comparatively light winds and fair weather
    found at the center of a severe tropical cyclone.

FAIRWAY - The parts of a waterway that are open and unobstructed for
    navigation. The main traveled part of a waterway; a marine
    thoroughfare.

FATHOM - A unit of measurement used for soundings. It is equal to 6 feet
    (1.83 meters).

FATHOMETER - The copyrighted trademark for a type of echo sounder.

FEEDER BEACH - An artificially widened beach serving to nourish downdrift beaches by natural littoral currents or forces.

FEEDER CURRENT - See CURRENT, FEEDER.

FEELING BOTTOM - The action of a deepwater wave on running into shoal water and beginning to be influenced by the bottom.

FETCH - The area in which SEAS are generated by a wind having a rather constant direction and speed. Sometimes used synonymously with FETCH LENGTH. Also GENERATING AREA.

FETCH LENGTH - The horizontal distance (in the direction of the wind) over which a wind generates SEAS or creates a WIND SETUP.

FIRTH - A narrow arm of the sea; also the opening of a river into the sea.

FIORD (FJORD) - A narrow, deep, steep-walled inlet of the sea, usually formed by entrance of the sea into a deep glacial trough.

FLOOD CURRENT - The tidal current toward shore or up a tidal stream, usually associated with the increase in the height of the tide.

FLOOD TIDE - The period of tide between low water and the succeeding high water; a rising tide. (See Figure A-10.)

FOAM LINE - The front of a wave as it advances shoreward, after it has broken. (See Figure A-4.)

FOLLOWING WIND - Generally, same as tailwind; in wave forecasting, wind blowing in the direction of ocean-wave advance.

FOREDUNE - The front dune immediately behind the backshore.

FORERUNNER - Low, long-period ocean SWELL which commonly precedes the main swell from a distant storm, especially a tropical cyclone.

FORESHORE - The part of the shore lying between the crest of the seaward berm (or upper limit of wave wash at high tide) and the ordinary low water mark, that is ordinarily traversed by the uprush and backrush of the waves as the tides rise and fall. See BEACH FACE. (See Figure A-1.)

FORWARD SPEED (HURRICANE) - Rate of movement (propagation) of the hurricane eye in mph or knots.

FREEBOARD - The additional height of a structure above design high water level to prevent overflow. Also, at a given time, the vertical distance between the water level and the top of the structure. On a ship, the distance from the waterline to main deck or gunwale.

FRINGING REEF - A coral reef attached directly to an insular or continental shore.

FRONT OF THE FETCH - In wave forecasting, the end of the generating area toward which the wind is blowing.

FROUDE NUMBER - The dimensionless ratio of the inertial force to the force of gravity for a given fluid flow. It may be given as $Fr = V/Lg$ where V is a characteristic velocity, L is a characteristic length, and g the acceleration of gravity; or as the square root of this number.

GENERATING AREA - In wave forecasting, the continuous area of water surface over which the wind blows in nearly a constant direction. Sometimes used synonymously with FETCH LENGTH. Also FETCH.

GENERATION OF WAVES - (1) The creation of waves by natural or mechanical means. (2) The creation and growth of waves caused by a wind blowing over a water surface for a certain period of time. The area involved is called the GENERATING AREA or FETCH.

GEOMETRIC MEAN DIAMETER - The diameter equivalent of the arithmetic mean of the logarithmic frequency distribution. In the analysis of beach sands, it is taken as that grain diameter determined graphically by the intersection of a straight line through selected boundary sizes, (generally points on the distribution curve where 16 and 84 percent of the sample is coarser by weight) and a vertical line through the median diameter of the sample.

GEOMETRIC SHADOW - In wave diffraction theory, the area outlined by drawing straight lines paralleling the direction of wave approach through the extremities of the protective structure. It differs from the actual protected area to the extent that the diffraction and refraction effects modify the wave pattern.

GEOMORPHOLOGY - That branch of both physiography and geology which deals with the form of the earth, the general configuration of its surface, and the changes that take place in the evolution of landform.

GRADIENT (GRADE) - See SLOPE. With reference to winds or currents, the rate of increase or decrease in speed, usually in the vertical; or the curve that represents this rate.

GRAVEL - See SOIL CLASSIFICATION.

GRAVITY WAVE - A wave whose velocity of propagation is controlled primarily by gravity. Water waves more than 2 inches long are considered gravity waves. Waves longer than 1 inch and shorter than 2 inches are in an indeterminate zone between CAPILLARY and GRAVITY WAVES. See RIPPLE.

GROIN (British, GROYNE) - A shore protection structure built (usually perpendicular to the shoreline) to trap littoral drift or retard erosion of the shore.

GROIN SYSTEM - A series of groins acting together to protect a section of beach. Commonly called a groin field.

GROUND SWELL - A long high ocean swell; also, this swell as it rises to prominent height in shallow water.

GROUND WATER - Subsurface water occupying the zone of saturation. In a strict sense, the term is applied only to water below the WATER TABLE.

GROUP VELOCITY - The velocity of a wave group. In deep water, it is equal to one-half the velocity of the individual waves within the group.

GULF - A large embayment in a coast; the entrance is generally wider than the length.

GUT - (1) A narrow passage such as a strait or inlet. (2) A channel in otherwise shallower water, generally formed by water in motion.

HALF-TIDE LEVEL - MEAN TIDE LEVEL.

HARBOR (British, HARBOUR) - Any protected water area affording a place of safety for vessels. See also PORT.

HARBOR OSCILLATION (Harbor Surging) - The nontidal vertical water movement in a harbor or bay. Usually the vertical motions are low, but when oscillations are excited by a tsunami or storm surge, they may be quite large. Variable winds, air oscillations, or surf beat also may cause oscillations. See SEICHE.

HEADLAND (HEAD) - A high steep-faced promontory extending into the sea.

HEAD OF RIP - The part of a rip current that has widened out seaward of the breakers. See also CURRENT, RIP; CURRENT, FEEDER; and NECK (RIP)

HEIGHT OF WAVE - See WAVE HEIGHT.

HIGH TIDE, HIGH WATER (HW) - The maximum elevation reached by each rising tide. See TIDE. (See Figure A-10.)

HIGH WATER OF ORDINARY SPRING TIDES (HWOST) - A tidal datum appearing in some British publications, based on high water of ordinary spring tides.

HIGHER HIGH WATER (HHW) - The higher of the two high waters of any tidal day. The single high water occurring daily during periods when the tide is diurnal is considered to be a higher high water. (See Figure A-10.)

HIGHER LOW WATER (HLW) - The higher of two low waters of any tidal day. (See Figure A-10.)

HIGH WATER - See HIGH TIDE.

HIGH WATER LINE - In strictness, the intersection of the plane of mean high water with the shore. The shoreline delineated on the nautical charts of the U. S. Coast and Geodetic Survey is an approximation of the high water line. For specific occurrences, the highest elevation on the shore reached during a storm or rising tide, including meteorological effects.

HINDCASTING, WAVE - The use of historic synoptic wind charts to calculate wave characteristics that probably occurred at some past time.

HOOK - A spit or narrow cape of sand or gravel which turns landward at the outer end.

HURRICANE - An intense tropical cyclone in which winds tend to spiral inward toward a core of low pressure, with maximum surface wind velocities that equal or exceed 75 mph (65 knots) for several minutes or longer at some points. TROPICAL STORM is the term applied if maximum winds are less than 75 mph.

HURRICANE PATH OR TRACK - Line of movement (propagation) of the eye through an area.

HURRICANE STAGE HYDROGRAPH - A continuous graph representing water level stages that would be recorded in a gage well located at a specified point of interest during the passage of a particular hurricane, assuming that effects of relatively short-period waves are eliminated from the record by damping features of the gage well. Unless specifically excluded and separately accounted for, hurricane surge hydrographs are assumed to include effects of astronomical tides, barometric pressure differences, and all other factors that influence water level stages within a properly designed gage well located at a specified point.

HURRICANE SURGE HYDROGRAPH - A continuous graph representing the difference between the hurricane stage hydrograph and the water stage hydrograph that would have prevailed at the same point and time if the hurricane had not occurred.

HURRICANE WIND PATTERN or ISOVEL PATTERNS - An actual or graphical representation of near-surface wind velocities covering the entire area of a hurricane at a particular instant. Isovels are lines connecting points of simultaneous equal wind velocities, usually referenced 30 feet above the surface, in knots or mph; wind directions at various points are indicated by arrows or deflection angles on the isovel charts. Isovel charts are usually prepared at each hour during a hurricane, but for each half hour during critical periods.

HYDRAULICALLY EQUIVALENT GRAINS - Sedimentary particles that settle at the same rate under the same conditions.

HYDROGRAPHY - (1) A configuration of an underwater surface including its relief, bottom materials, coastal structures, etc. (2) The description and study of seas, lakes, rivers, and other waters.

HYPOTHETICAL HURRICANE ("HYPO-HURRICANE") - A representation of a hurricane, with specified characteristics, that is assumed to occur in a particular study area, following a specified path and timing sequence.

    TRANSPOSED - A hypo-hurricane based on the storm transposition principle is assumed to have wind patterns and other characteristics basically comparable to a specified hurricane of record, but is *transposed* to follow a new path to serve as a basis for computing a hurricane surge hydrograph that would be expected at a selected point. Moderate adjustments in timing or rate of forward movement may be made also, if these are compatible with meteorological considerations and study objectives.

    HYPO-HURRICANE BASED ON GENERALIZED PARAMETERS - Hypo-hurricane estimates based on various logical combinations of hurricane characteristics used in estimating hurricane surge magnitudes corresponding to a range of probabilities and potentialities. The *Standard Project Hurricane* (SPH) is most commonly used for this purpose, but estimates corresponding to more severe or less severe assumptions are important in some project investigations.

    STANDARD PROJECT HURRICANE (SPH) - A hypothetical hurricane intended to represent the most severe combination of hurricane parameters that is *reasonably characteristic* of a specified region, excluding extremely rare combinations. It is further assumed that the SPH would approach a given project site from such direction, and at such rate of movement as to produce the highest hurricane surge hydrograph, considering pertinent hydraulic characteristics of the area. Based on this concept, and extensive meteorological studies and probability analyses, a tabulation of "Standard Project Hurricane Index Characteristics" mutually agreed upon by representatives of the U. S. Weather Bureau and the Corps of Engineers, is available.

PROBABLE MAXIMUM HURRICANE - A hypo-hurricane that might result from the most severe combination of hurricane parameters that is considered reasonably possible in the region involved, if the hurricane should approach the point under study along a critical path and at optimum rate of movement. This estimate is substantially more severe than the SPH criteria.

DESIGN HURRICANE - A representation of a hurricane with specified characteristics that would produce hurricane surge hydrographs and coincident wave effects at various key locations along a proposed project alinement. It governs the project design after economics and other factors have been duly considered. The design hurricane may be more or less severe than the SPH, depending on economics, risk, and local considerations.

IMPERMEABLE GROIN - A groin through which sand cannot pass.

INDIAN SPRING LOW WATER - The approximate level of the mean of lower low waters at spring tides, used principally in the Indian Ocean and along the east coast of Asia. Also INDIAN TIDE PLANE.

INDIAN TIDE PLANE - The datum of INDIAN SPRING LOW WATER.

INLET - (1) A short, narrow waterway connecting a bay, lagoon, or similar body of water with a large parent body of water. (2) An arm of the sea (or other body of water), that is long compared to its width, and may extend a considerable distance inland. See also TIDAL INLET.

INLET GORGE - Generally, the deepest region of an inlet channel.

INSHORE (ZONE) - In beach terminology, the zone of variable width extending from the low water line through the breaker zone. SHOREFACE. (See Figure A-1.)

INSHORE CURRENT - Any current in or landward of the breaker zone.

INSULAR SHELF - The zone surrounding an island extending from the low water line to the depth (usually about 100 fathoms) where there is a marked or rather steep descent toward the great depths.

INTERNAL WAVES - Waves that occur within a fluid whose density changes with depth, either abruptly at a sharp surface of discontinuity (an interface) or gradually. Their amplitude is greatest at the density discontinuity or, in the case of a gradual density change, somewhere in the interior of the fluid and not at the free upper surface where the surface waves have their maximum amplitude.

IRROTATIONAL WAVE - A wave with fluid particles that do not revolve around an axis through their centers, although the particles themselves may travel in circular or nearly circular orbits. Irrotational waves may be progressive, standing, oscillatory, or translatory. For example, the Airy, Stokes, cnoidal and solitary wave theories describe irrotational waves. See TROCHOIDAL WAVE.

ISOBATH - A contour line connecting points of equal water depths on a chart.

ISOVEL PATTERN - See HURRICANE WIND PATTERN.

ISTHMUS - A narrow strip of land, bordered on both sides by water, that connects two larger bodies of land.

JET - To place (as a pile, slab, or pipe) in the ground by means of a jet of water acting at the lower end.

JETTY - (1) (U. S. usage) On open seacoasts, a structure extending into a body of water, and designed to prevent shoaling of a channel by littoral materials, and to direct and confine the stream or tidal flow. Jetties are built at the mouth of a river or tidal inlet to help deepen and stabilize a channel. (2) (British usage) Jetty is synonymous with "wharf" or "pier". See TRAINING WALL.

KEY - A low insular bank of sand, coral, etc., as one of the islets off the southern coast of Florida, also CAY.

KINETIC ENERGY (OF WAVES) - In a progressive oscillatory wave, a summation of the energy of motion of the particles within the wave.

KNOLL - A submerged elevation of rounded shape rising less than 1,000 meters from the ocean floor, and of limited extent across the summit. See SEAMOUNT.

KNOT - The unit of speed used in navigation. It is equal to 1 nautical mile (6,076.115 feet or 1,852 meters) per hour.

LAGGING - See DAILY RETARDATION (OF TIDES).

LAGOON - A shallow body of water, as a pond or lake, usually connected to the sea. (See Figures A-8 and A-9.)

LAND BREEZE - A light wind blowing from the land to the sea caused by unequal cooling of land and water masses.

LAND-SEA BREEZE - The combination of a land breeze and a sea breeze as a diurnal phenomenon.

LANDLOCKED - An area of water enclosed, or nearly enclosed, by land, as a bay or a harbor (thus, protected from the sea).

LANDMARK - A conspicuous object natural or artificial, located near or on land which aids in fixing the position of an observer.

LEAD LINE - A line, wire, or cord used in sounding. It is weighted at one end with a plummet (sounding lead). Also SOUNDING LINE.

LEE - (1) Shelter, or the part or side sheltered or turned away from the wind or waves. (2) (Chiefly nautical) The quarter or region toward which the wind blows.

LEEWARD - The direction *toward* which the wind is blowing; the direction toward which waves are traveling.

LENGTH OF WAVE - The horizontal distance between similar points on two successive waves measured perpendicularly to the crest. (See Figure A-3.)

LEVEE - A dike or embankment to protect land from inundation.

LIMIT OF BACKRUSH, LIMIT OF BACKWASH - See BACKWASH.

LITTORAL - Of or pertaining to a shore, especially of the sea.

LITTORAL CURRENT - See CURRENT, LITTORAL.

LITTORAL DEPOSITS - Deposits of littoral drift.

LITTORAL DRIFT - The sedimentary *material* moved in the littoral zone under the influence of waves and currents.

LITTORAL TRANSPORT - The *movement* of littoral drift in the littoral zone by waves and currents. Includes movement parallel (longshore transport) and perpendicular (on-offshore transport) to the shore.

LITTORAL TRANSPORT RATE - Rate of transport of sedimentary material parallel to or perpendicular to the shore in the littoral zone. Usually expressed in cubic yards (meters) per year. Commonly used as synonymous with LONGSHORE TRANSPORT RATE.

LITTORAL ZONE - In beach terminology, an indefinite zone extending seaward from the shoreline to just beyond the breaker zone.

LOAD - The quantity of sediment transported by a current. It includes the suspended load of small particles, and the bedload of large particles that move along the bottom.

LONGSHORE - Parallel to and near the shoreline.

LONGSHORE BAR - A bar running roughly parallel to the shoreline.

LONGSHORE CURRENT - See CURRENT, LONGSHORE.

LONGSHORE TRANSPORT RATE - Rate of transport of sedimentary material parallel to the shore. Usually expressed in cubic yards (meters) per year. Commonly used as synonymous with LITTORAL TRANSPORT RATE.

LOOP - That part of a STANDING WAVE where the vertical motion is greatest and the horizontal velocities are least. LOOPS (sometimes called ANTINODES) are associated with CLAPOTIS, and with SEICHE action resulting from wave reflections. (See also NODE.)

LOWER HIGH WATER (LHW) - The lower of the two high waters of any tidal day. (See Figure A-10.)

LOWER LOW WATER (LLW) - The lower of the two low waters of any tidal day. The single low water occurring daily during periods when the tide is diurnal is considered to be a lower low water. (See Figure A-10.)

LOW TIDE (LOW WATER, LW) - The minimum elevation reached by each falling tide. See TIDE. (See Figure A-10.)

LOW WATER DATUM - An approximation to the plane of mean low water that has been adopted as a standard reference plane. See also DATUM PLANE and CHART DATUM.

LOW WATER LINE - The intersection of any standard low tide datum plane with the shore.

LOW WATER OF ORDINARY SPRING TIDES (LWOST) - A tidal datum appearing in some British publications, based on low water of ordinary spring tides.

MANGROVE - A tropical tree with interlacing prop roots, confined to low-lying brackish areas.

MARIGRAM - A graphic record of the rise and fall of the tide.

MARSH - An area of soft, wet, or periodically inundated land, generally treeless and usually characterized by grasses and other low growth.

MARSH, SALT - A marsh periodically flooded by salt water.

MASS TRANSPORT - The net transfer of water by wave action in the direction of wave travel. See ORBIT.

MEAN DIAMETER, GEOMETRIC - See GEOMETRIC MEAN DIAMETER.

MEAN HIGHER HIGH WATER (MHHW) - The average height of the higher high waters over a 19-year period. For shorter periods of observation, corrections are applied to eliminate known variations and reduce the result to the equivalent of a mean 19-year value.

MEAN HIGH WATER (MHW) - The average height of the high waters over a 19-year period. For shorter periods of observations, corrections are applied to eliminate known variations and reduce the results to the equivalent of a mean 19-year value. All high water heights are included in the average where the type of tide is either semidiurnal or mixed. Only the higher high water heights are included in the average where the type of tide is diurnal. So determined, mean high water in the latter case is the same as mean higher high water.

MEAN HIGH WATER SPRINGS - The average height of the high waters occurring at the time of spring tide. Frequently abbreviated to HIGH WATER SPRINGS.

MEAN LOWER LOW WATER (MLLW) - The average height of the lower low waters over a 19-year period. For shorter periods of observations, corrections are applied to eliminate known variations and reduce the results to the equivalent of a mean 19-year value. Frequently abbreviated to LOWER LOW WATER.

MEAN LOW WATER (MLW) - The average height of the low waters over a 19-year period. For shorter periods of observations, corrections are applied to eliminate known variations and reduce the results to the equivalent of a mean 19-year value. All low water heights are included in the average where the type of tide is either semidiurnal or mixed. Only lower low water heights are included in the average where the type of tide is diurnal. So determined, mean low water in the latter case is the same as mean lower low water.

MEAN LOW WATER SPRINGS - The average height of low waters occurring at the time of the spring tides. It is usually derived by taking a plane depressed below the half-tide level by an amount equal to one-half the spring range of tide, necessary corrections being applied to reduce the result to a mean value. This plane is used to a considerable extent for hydrographic work outside of the United States and is the plane of reference for the Pacific approaches to the Panama Canal. Frequently abbreviated to LOW WATER SPRINGS.

MEAN SEA LEVEL - The average height of the surface of the sea for all stages of the tide over a 19-year period, usually determined from hourly height readings. Not necessarily equal to MEAN TIDE LEVEL.

MEAN TIDE LEVEL - A plane midway between MEAN HIGH WATER AND MEAN LOW WATER. Not necessarily equal to MEAN SEA LEVEL. Also called HALF-TIDE LEVEL.

MEDIAN DIAMETER - The diameter which marks the division of a given sand sample into two equal parts by weight, one part containing all grains larger than that diameter and the other part containing all grains smaller.

MEGARIPPLE - See SAND WAVE.

MIDDLE-GROUND SHOAL - A shoal formed by ebb and flood tides in the middle of the channel of the lagoon or estuary end of an inlet.

MINIMUM DURATION - See DURATION, MINIMUM.

MINIMUM FETCH - The least distance in which steady state wave conditions will develop for a wind of given speed blowing a given duration of time.

MIXED TIDE - A type of tide in which the presence of a diurnal wave is conspicuous by a large inequality in either the high- or low-water heights with two high waters and two low waters usually occurring each tidal day. In strictness, all tides are mixed, but the name is usually applied without definite limits to the tide intermediate to those predominantly semidiurnal and those predominantly diurnal. (See Figure A-10.)

MOLE - In coastal terminology, a massive land-connected, solid-fill structure of earth (generally revetted), masonry, or large stone. It may serve as a breakwater or pier.

MONOCHROMATIC WAVES - A series of waves generated in a laboratory; each wave has the same length and period.

MONOLITHIC - Like a single stone or block. In coastal structures, the type of construction in which the structure's component parts are bound together to act as one.

MUD - A fluid-to-plastic mixture of finely divided particles of solid material and water.

NAUTICAL MILE - The length of a minute of arc, 1/21,600 of an average great circle of the earth. Generally one minute of latitude is considered equal to one nautical mile. The accepted United States value as of 1 July 1959 is 6,076.115 feet or 1,852 meters, approximately 1.15 times as long as the statute mile of 5,280 feet. Also geographical mile.

NEAP TIDE - A tide occurring near the time of quadrature of the moon with the sun. The neap tidal range is usually 10 to 30 percent less than the mean tidal range.

NEARSHORE (ZONE) - In beach terminology an indefinite zone extending seaward from the shoreline well beyond the breaker zone. It defines the area of NEARSHORE CURRENTS. (See Figure A-1.)

NEARSHORE CIRCULATION - The ocean circulation pattern composed of the CURRENTS, NEARSHORE and CURRENTS, COASTAL. See CURRENT.

NEARSHORE CURRENT SYSTEM - The current system caused primarily by wave action in and near the breaker zone, and which consists of four parts: The shoreward mass transport of water; longshore currents; seaward return flow, including rip currents; and the longshore movement of the expanding heads of rip currents. (See Figure A-7.) See also NEARSHORE CIRCULATION.

NECK - (1) The narrow band of water flowing seaward through the surf. Also RIP. (2) The narrow strip of land connecting two larger bodies of land, as an isthmus.

NIP - The cut made by waves in a shoreline of emergence.

NODAL ZONE - An area in which the predominant direction of the LONGSHORE
    TRANSPORT changes.

NODE - That part of a STANDING WAVE where the vertical motion is least
    and the horizontal velocities are greatest. Nodes are associated
    with CLAPOTIS and with SEICHE action resulting from wave reflections.
    See also LOOP.

NOURISHMENT - The process of replenishing a beach. It may be brought
    about naturally, by longshore transport, or artificially by the
    deposition of dredged materials.

OCEANOGRAPHY - The study of the sea, embracing and indicating all
    knowledge pertaining to the sea's physical boundaries, the chem-
    istry and physics of sea water, and marine biology.

OFFSHORE - (1) In beach terminology, the comparatively flat zone of
    variable width, extending from the breaker zone to the seaward edge
    of the Continental Shelf. (2) A direction seaward from the shore.
    (See Figure A-1.)

OFFSHORE BARRIER - See BARRIER BEACH.

OFFSHORE CURRENT - (1) Any current in the offshore zone. (2) Any
    current flowing away from shore.

OFFSHORE WIND - A wind blowing seaward from the land in the coastal area.

ONSHORE - A direction landward from the sea.

ONSHORE WIND - A wind blowing landward from the sea in the coastal area.

OPPOSING WIND - In wave forecasting, a wind blowing in a direction opposite
    to the ocean-wave advance; generally, same as headwind.

ORBIT - In water waves, the path of a water particle affected by the wave
    motion. In deepwater waves the orbit is nearly circular and in
    shallow-water waves the orbit is nearly elliptical. In general,
    the orbits are slightly open in the direction of wave motion giving
    rise to MASS TRANSPORT. (See Figure A-3.)

ORBITAL CURRENT - The flow of water accompanying the orbital movement of
    the water particles in a wave. Not to be confused with wave-generated
    LITTORAL CURRENTS. (See Figure A-3.)

ORTHOGONAL - On a wave-refraction diagram, a line drawn perpendicularly to
    the wave crests. (See Figure A-6.)

OSCILLATION - A periodic motion backward and forward. To vibrate or vary
    above and below a mean value.

A-24

OSCILLATORY WAVE - A wave in which each individual particle oscillates about a point with little or no permanent change in mean position. The term is commonly applied to progressive oscillatory waves in which only the form advances, the individual particles moving in closed or nearly closed orbits. Distinguished from a WAVE OF TRANSLATION. See also ORBIT.

OUTFALL - A structure extending into a body of water for the purpose of discharging sewage, storm runoff, or cooling water.

OVERTOPPING - Passing of water over the top of a structure as a result of wave runup or surge action.

OVERWASH - That portion of the uprush that carries over the crest of a berm or of a structure.

PARAPET - A low wall built along the edge of a structure as on a seawall or quay.

PARTICLE VELOCITY - The velocity induced by wave motion with which a specific water particle moves within a wave.

PASS - In hydrographic usage, a navigable channel through a bar, reef, or shoal, or between closely adjacent islands.

PEBBLES - See SOIL CLASSIFICATION.

PENINSULA - An elongated body of land nearly surrounded by water, and connected to a larger body of land.

PERCHED BEACH - A beach or fillet of sand retained above the otherwise normal profile level by a submerged dike.

PERCOLATION - The process by which water flows through the interstices of a sediment. Specifically, in wave phenomena, the process by which wave action forces water through the interstices of the bottom sediment. Tends to reduce wave heights.

PERIODIC CURRENT - A current caused by the tide-producing forces of the moon and the sun, a part of the same general movement of the sea that is manifested in the vertical rise and fall of the tides. See also CURRENT, FLOOD and CURRENT, EBB.

PERMANENT CURRENT - A current that runs continuously, independent of the tides and temporary causes. Permanent currents include the fresh-water discharge of a river and the currents that form the general circulatory systems of the oceans.

PERMEABLE GROIN - A groin with openings large enough to permit passage of appreciable quantities of littoral drift.

PETROGRAPHY - The systematic description and classification of rocks.

PHASE - In surface wave motion, a point in the period to which the wave motion has advanced with respect to a given initial reference point.

PHASE INEQUALITY - Variations in the tides or tidal currents associated with changes in the phase of the moon in relation to the sun.

PHASE VELOCITY - Propagation velocity of an individual wave as opposed to the velocity of a wave group.

PHI GRADE SCALE - A logarithmic transformation of the Wentworth grade scale for size classifications of sediment grains based on the negative logarithm to the base 2 of the particle diameter. $\phi = -\log_2 d$. See SOIL CLASSIFICATION.

PIER - A structure, usually of open construction, extending out into the water from the shore, to serve as a landing place, a recreational facility, etc., rather than to afford coastal protection. In the Great Lakes, a term sometimes improperly applied to jetties.

PILE - A long, heavy timber or section of concrete or metal to be driven or jetted into the earth or seabed to serve as a support or protection.

PILE, SHEET - A pile with a generally slender flat cross section to be driven into the ground or seabed and meshed or interlocked with like members to form a diaphragm, wall, or bulkhead.

PILING - A group of piles.

PLAIN, COASTAL - See COASTAL PLAIN.

PLANFORM - The outline or shape of a body of water as determined by the stillwater line.

PLATEAU - A land area (usually extensive) having a relatively level surface raised sharply above adjacent land on at least one side; table land. A similar undersea feature.

PLUNGE POINT - (1) For a plunging wave, the point at which the wave curls over and falls. (2) The final breaking point of the waves just before they rush up on the beach. (See Figure A-1.)

PLUNGING BREAKER - See BREAKER.

POCKET BEACH - A beach, usually small, in a coastal reentrant or between two littoral barriers.

POINT - The extreme end of a cape, or the outer end of any land area protruding into the water, usually less prominent than a cape.

PORT - A place where vessels may discharge or receive cargo; may be the entire harbor including its approaches and anchorages, or may be the commercial part of a harbor where the quays, wharves, facilities for transfer of cargo, docks, and repair shops are situated.

POTENTIAL ENERGY OF WAVES - In a progressive oscillatory wave, the energy resulting from the elevation or depression of the water surface from the undisturbed level.

PRISM - See TIDAL PRISM.

PROBABLE MAXIMUM WATER LEVEL - A hypothetical water level (exclusive of wave runup from normal wind-generated waves) that might result from the most severe combination of hydrometeorological, geoseismic and other geophysical factors that is considered reasonably possible in the region involved, with each of these factors considered as affecting the locality in a maximum manner.

This level represents the physical response of a body of water to maximum applied phenomena such as hurricanes, moving squall lines, other cyclonic meteorological events, tsunamis, and astronomical tide combined with maximum probable ambient hydrological conditions such as wave setup, rainfall, runoff, and river flow. It is a water level with virtually no risk of being exceeded.

PROFILE, BEACH - The intersection of the ground surface with a vertical plane; may extend from the top of the dune line to the seaward limit of sand movement. (See Figure A-1.)

PROGRESSION (of a beach) - See ADVANCE.

PROGRESSIVE WAVE - A wave that moves relative to a fixed coordinate system in a fluid. The direction in which it moves is termed the direction of wave propagation.

PROMONTORY - A high point of land projecting into a body of water; a HEADLAND.

PROPAGATION OF WAVES - The transmission of waves through water.

PROTOTYPE - In laboratory usage, the full-scale structure, concept, or phenomenon used as a basis for constructing a scale model or copy.

QUAY (Pronounced KEY) - A stretch of paved bank, or a solid artificial landing place parallel to the navigable waterway, for use in loading and unloading vessels.

QUICKSAND - Loose, yielding, wet sand which offers no support to heavy objects. The upward flow of the water has a velocity that eliminates contact pressures between the sand grains, and causes the sand-water mass to behave like a fluid.

RADIUS OF MAXIMUM WINDS - Distance from the eye of a hurricane, where surface and wind velocities are zero to the place where surface wind speeds are maximum.

RAY, WAVE - See ORTHOGONAL.

RECESSION (of a beach) - (1) A continuing landward movement of the shoreline. (2) A net landward movement of the shoreline over a specified time. Also RETROGRESSION.

REEF - An offshore consolidated rock hazard to navigation with a least depth of 10 fathoms (about 20 meters) or less.

REEF, ATOLL - See ATOLL.

REEF, BARRIER - See BARRIER REEF.

REEF, FRINGING - See FRINGING REEF.

REEF, SAND - Synonymous with BAR.

REFERENCE PLANE - See DATUM PLANE.

REFERENCE STATION - A place for which tidal constants have previously been determined and which is used as a standard for the comparison of simultaneous observations at a second station; also a station for which independent daily predictions are given in the tide or current tables from which corresponding predictions are obtained for other stations by means of differences or factors.

REFLECTED WAVE - That part of an incident wave that is returned seaward when a wave impinges on a steep beach, barrier, or other reflecting surface.

REFRACTION (OF WATER WAVES) - (1) The process by which the direction of a wave moving in shallow water at an angle to the contours is changed. The part of the wave advancing in shallower water moves more slowly than that part still advancing in deeper water, causing the wave crest to bend toward alignment with the underwater contours. (2) The bending of wave crests by currents. (See Figure A-5.)

REFRACTION COEFFICIENT - The square root of the ratio of the spacing between adjacent orthogonals in deep water and in shallow water at a selected point. When multiplied by the SHOALING FACTOR and a factor for friction and percolation, this becomes the WAVE HEIGHT COEFFICIENT or the ratio of the refracted wave height at any point to the deepwater wave height. Also the square root of the ENERGY COEFFICIENT.

REFRACTION DIAGRAM - A drawing showing positions of wave crests and/or orthogonals in a given area for a specific deepwater wave period and direction. (See Figure A-6.)

RESONANCE - The phenomenon of amplification of a free wave or oscillation of a system by a forced wave or oscillation of exactly equal period. The forced wave may arise from an impressed force upon the system or from a boundary condition.

RETARDATION - The amount of time by which corresponding tidal phases grow later day by day (about 50 minutes).

RETROGRESSION OF A BEACH - (1) A continuing landward movement of the shoreline. (2) A net landward movement of the shoreline over a specified time. Also RECESSION.

REVETMENT - A facing of stone, concrete, etc., built to protect a scarp, embankment, or shore structure against erosion by wave action or currents.

REYNOLDS NUMBER - The dimensionless ratio of the inertial force to the viscous force in fluid motion,

$$R_e = \frac{LV}{\nu}$$

where L is a characteristic length, ν the kinematic viscosity, and V a characteristic velocity. The Reynolds number is of importance in the theory of hydrodynamic stability and the origin of turbulence.

RIA - A long, narrow inlet, with depth gradually diminishing inward.

RIDGE, BEACH - A nearly continuous mound of beach material that has been shaped up by wave or other action. Ridges may occur singly or as a series of approximately parallel deposits. (See Figure A-7.) British usage, fulls.

RILL MARKS - Tiny drainage channels in a beach caused by the flow seaward of water left in the sands of the upper part of the beach after the retreat of the tide or after the dying down of storm waves.

RIP - A body of water made rough by waves meeting an opposing current, particularly a tidal current; often found where tidal currents are converging and sinking.

RIPARIAN - Pertaining to the banks of a body of water.

RIPARIAN RIGHTS - The rights of a person owning land containing or bordering on a watercourse or other body of water in or to its banks, bed or waters.

RIP CURRENT - A strong surface current flowing seaward from the shore. It usually appears as a visible band of agitated water and is the return movement of water piled up on the shore by incoming waves and wind. With the seaward movement concentrated in a limited band its velocity is somewhat accentuated. A rip consists of three parts: the FEEDER CURRENTS flowing parallel to the shore inside the breakers; the NECK, where the feeder currents converge and flow through the breakers in a narrow band or "rip"; and the HEAD, where the current widens and slackens outside the breaker line. A rip current is often miscalled a rip tide. Also RIP SURF. See NEARSHORE CURRENT SYSTEM. (See Figure A-7.)

RIP SURF - See RIP CURRENT.

RIPPLE - (1) The ruffling of the surface of water, hence a little curling wave or undulation. (2) A wave less than 2 inches long controlled to a significant degree by both surface tension and gravity. See WAVE, CAPILLARY and WAVE, GRAVITY.

RIPPLES (BED FORMS) - Small bed forms with wavelengths less than 1 foot and heights less than 0.1 foot.

RIPRAP - A layer, facing, or protective mound of stones randomly placed to prevent erosion, scour, or sloughing of a structure or embankment; also the stone so used.

ROADSTEAD (Nautical) - A sheltered area of water near shore where vessels may anchor in relative safety. Also road.

ROLLER - An indefinite term, sometimes considered to denote one of a series of long-crested, large waves which roll in on a shore, as after a storm.

RUBBLE - (1) Loose angular waterworn stones along a beach. (2) Rough, irregular fragments of broken rock.

RUBBLE-MOUND STRUCTURE - A mound of random-shaped and random-placed stones protected with a cover layer of selected stones or specially shaped concrete armor units. (Armor units in primary cover layer may be placed in orderly manner or dumped at random.)

RUNNEL - A corrugation or trough formed in the foreshore or in the bottom just offshore by waves or tidal currents.

RUNUP - The rush of water up a structure or beach on the breaking of a wave. Also UPRUSH. The amount of runup is the vertical height above stillwater level that the rush of water reaches.

SALTATION - That method of sand movement in a fluid in which individual particles leave the bed by bounding nearly vertically and, because the motion of the fluid is not strong or turbulent enough to retain them in suspension, return to the bed at some distance downstream. The travel path of the particles is a series of hops and bounds.

SALT MARSH - A marsh periodically flooded by salt water.

SAND - See SOIL CLASSIFICATION.

SANDBAR - (1) See BAR. (2) In a river, a ridge of sand built up to or near the surface by river currents.

SAND BYPASSING - See BYPASSING, SAND.

SAND REEF - Synonymous with BAR.

SAND WAVE (or MEGARIPPLE) - A large wavelike sediment feature composed of sand in very shallow water. Wavelength may reach 100 meters; amplitude is about 0.5 meters.

SCARP - See ESCARPMENT.

SCARP, BEACH - An almost vertical slope along the beach caused by erosion by wave action. It may vary in height from a few inches to several feet, depending on wave action and the nature and composition of the beach. (See Figure A-1.)

SCOUR - Removal of underwater material by waves and currents, especially at the base or toe of a shore structure.

SEAS - Waves caused by wind at the place and time of observation.

SEA STATE - Description of the sea surface with regard to wave action. Also called state of sea.

SEA BREEZE - A light wind blowing from the sea toward the land caused by unequal heating of land and water masses.

SEA CLIFF - A cliff situated at the seaward edge of the coast.

SEA LEVEL - See MEAN SEA LEVEL.

SEAMOUNT - An elevation rising more than 1,000 meters above the ocean floor, and of limited extent across the summit.

SEA PUSS - A dangerous longshore current; a rip current caused by return flow; loosely, the submerged channel or inlet through a bar caused by those currents.

SEASHORE - The SHORE of a sea or ocean.

SEAWALL - A structure separating land and water areas, primarily
designed to prevent erosion and other damage due to wave action.
See also BULKHEAD.

SEICHE - (1) A standing wave oscillation of an enclosed water body that
continues, pendulum fashion, after the cessation of the originating
force, which may have been either seismic or atmospheric. (2) An
oscillation of a fluid body in response to a disturbing force having
the same frequency as the natural frequency of the fluid system.
Tides are now considered to be seiches induced primarily by the
periodic forces caused by the sun and moon. (3) In the Great Lakes
area, any sudden rise in the water of a harbor or a lake whether or
not it is oscillatory. Although inaccurate in a strict sense, this
usage is well established in the Great Lakes area.

SEISMIC SEA WAVE (TSUNAMI) - A long-period wave caused by an underwater
seismic disturbance or volcanic eruption. Commonly misnamed
"tidal wave".

SEMIDIURNAL TIDE - A tide with two high and two low waters in a tidal
day with comparatively little diurnal inequality (See Figure A-10.)

SET OF CURRENT - The direction toward which a current flows.

SETUP, WAVE - Superelevation of the water surface over normal surge
elevation due to onshore mass transport of the water by wave
action alone.

SETUP, WIND - See WIND SETUP.

SHALLOW WATER - (1) Commonly, water of such a depth that surface waves
are noticeably affected by bottom topography. It is customary to
consider water of depths less than one-half the surface wavelength
as shallow water. See TRANSITIONAL ZONE and DEEP WATER. (2) More
strictly, in hydrodynamics with regard to progressive gravity waves,
water in which the depth is less than 1/25 the wavelength. Also
called VERY SHALLOW WATER.

SHEET PILE - See PILE, SHEET.

SHELF, CONTINENTAL - See CONTINENTAL SHELF.

SHELF, INSULAR - See INSULAR SHELF.

SHINGLE - (1) Loosely and commonly, any beach material coarser than
ordinary gravel, especially any having flat or flattish pebbles.
(2) Strictly and accurately, beach material of smooth, well-rounded
pebbles that are roughly the same size. The spaces between pebbles
are not filled with finer materials. Shingle often gives out a
musical sound when stepped on.

SHOAL (noun) - A detached elevation of the sea bottom, comprised of any material except rock or coral, which may endanger surface navigation.

SHOAL (verb) - (1) To *become* shallow gradually. (2) To *cause* to become shallow. (3) To *proceed* from a greater to a lesser depth of water.

SHOALING COEFFICIENT - The ratio of the height of a wave in water of any depth to its height in deep water with the effects of refraction, friction, and percolation eliminated. Sometimes SHOALING FACTOR or DEPTH FACTOR. See also ENERGY COEFFICIENT and REFRACTION COEFFICIENT.

SHORE - The narrow strip of land in immediate contact with the sea, including the zone between high and low water lines. A shore of unconsolidated material is usually called a beach. (See Figure A-1.)

SHOREFACE - The narrow zone seaward from the low tide SHORELINE covered by water over which the beach sands and gravels actively oscillate with changing wave conditions. See INSHORE (ZONE) and Figure A-1.

SHORELINE - The intersection of a specified plane of water with the shore or beach. (e.g., the highwater shoreline would be the intersection of the plane of mean high water with the shore or beach.) The line delineating the shoreline on U. S. Coast and Geodetic Survey nautical charts and surveys approximates the mean high water line.

SIGNIFICANT WAVE - A statistical term relating to the one-third highest waves of a given wave group and defined by the average of their heights and periods. The composition of the higher waves depends upon the extent to which the lower waves are considered. Experience indicates that a careful observer who attempts to establish the character of the higher waves will record values which approximately fit the definition of the significant wave.

SIGNIFICANT WAVE HEIGHT - The average height of the one-third highest waves of a given wave group. Note that the composition of the highest waves depends upon the extent to which the lower waves are considered. In wave record analysis, the average height of the highest one-third of a selected number of waves, this number being determined by dividing the time of record by the significant period. Also CHARACTERISTIC WAVE HEIGHT.

SIGNIFICANT WAVE PERIOD - An arbitrary period generally taken as the period of the one-third highest waves within a given group. Note that the composition of the highest waves depends upon the extent to which the lower waves are considered. In wave record analysis, this is determined as the average period of the most frequently recurring of the larger well-defined waves in the record under study.

SILT - See SOIL CLASSIFICATION.

SINUSOIDAL WAVE - An oscillatory wave having the form of a sinusoid.

SLACK TIDE (SLACK WATER) - The state of a tidal current when its velocity is near zero, especially the moment when a reversing current changes direction and its velocity is zero. Sometimes considered the intermediate period between ebb and flood currents during which the velocity of the currents is less than 0.1 knot. See STAND OF TIDE.

SLIP - A berthing space between two piers.

SLOPE - The degree of inclination to the horizontal. Usually expressed as a ratio, such as 1:25 or 1 on 25, indicating 1 unit vertical rise in 25 units of horizontal distance; or in a decimal fraction (0.04); degrees (2° 18'); or percent (4%).

SLOUGH - See BAYOU.

SOIL CLASSIFICATION (size) - An arbitrary division of a continuous scale of grain sizes such that each scale unit or grade may serve as a convenient class interval for conducting the analysis or for expressing the results of an analysis. There are many classifications used; the two most often used are shown graphically on the next page.

SOLITARY WAVE - A wave consisting of a single elevation (above the original water surface), its height not necessarily small compared to the depth, and neither followed nor preceded by another elevation or depression of the water surfaces.

SORTING COEFFICIENT - A coefficient used in describing the distribution of grain sizes in a sample of unconsolidated material. It is defined as $S_0 = \sqrt{Q_1/Q_3}$, where Q_1 is the diameter (in millimeters) which has 75 percent of the cumulative size-frequency (by weight) distribution smaller than itself and 25 percent larger than itself, and Q_3 is that diameter having 25 percent smaller and 75 percent larger than itself.

SOUND (noun) - (1) A wide waterway between the mainland and an island, or a wide waterway connecting two sea areas. See also STRAIT. (2) A relatively long arm of the sea or ocean forming a channel between an island and a mainland or connecting two larger bodies, as a sea and the ocean, or two parts of the same body; usually wider and more extensive than a strait.

SOUND (verb) - To measure the depth of the water.

SOUNDING - A measured depth of water. On hydrographic charts the soundings are adjusted to a specific plane of reference (SOUNDING DATUM).

SOUNDING DATUM - The plane to which soundings are referred. See also CHART DATUM.

GRAIN SIZE SCALES
(Soil Classification)

Wentworth Scale (Size Description)		Phi Units φ *	Grain Diameter, D (mm)	U.S. Std. Sieve Size	Unified Soil Classification (USC)	
Boulder		-8	256		Cobble	
Cobble			76.2	3"		
		-6	64.0		Coarse	Gravel
			19.0	3/4"		
Pebble			4.76	No. 4	Fine	
		-2	4.0		Coarse	Sand
Granule		-1	2.0	No. 10		
Sand	Very Coarse	0	1.0		Medium	
	Coarse	1	0.5			
	Medium		0.42	No. 40		
		2	0.25		Fine	
	Fine	3	0.125			
	Very Fine		0.074	No. 200		
		4	0.0625		Silt or Clay	
Silt		8	0.00391			
Clay		12	0.00024			
Colloid						

* $\phi = -\log_2 D$ (mm)

SOUNDING LINE - A line, wire, or cord used in sounding. It is weighted at one end with a plummet (sounding lead). Also LEADLINE.

SPILLING BREAKER - See BREAKER.

SPIT - A small point of land or a narrow shoal projecting into a body of water from the shore. (See Figure A-9.)

SPRING TIDE - A tide that occurs at or near the time of new or full moon (syzygy), and which rises highest and falls lowest from the mean sea level.

STANDARD PROJECT HURRICANE - See HYPOTHETICAL HURRICANE.

STAND OF TIDE - An interval at high or low water when there is no sensible change in the height of the tide. The water level is stationary at high and low water for only an instant, but the change in level near these times is so slow that it is not usually perceptible. See SLACK TIDE.

STANDING WAVE - A type of wave in which the surface of the water oscillates vertically between fixed points, called nodes, without progression. The points of maximum vertical rise and fall are called antinodes or loops. At the nodes, the underlying water particles exhibit no vertical motion, but maximum horizontal motion. At the antinodes, the underlying water particles have no horizontal motion but maximum vertical motion. They may be the result of two equal progressive wave trains traveling through each other in opposite directions. Sometimes called CLAPOTIS or STATIONARY WAVE.

STATIONARY WAVE - A wave of essentially stable form which does not move with respect to a selected reference point; a fixed swelling. Sometimes called STANDING WAVE.

STILLWATER LEVEL - The elevation that the surface of the water would assume if all wave action were absent.

STOCKPILE - Sand piled on a beach foreshore to nourish downdrift beaches by natural littoral currents or forces. See FEEDER BEACH.

STONE, DERRICK - Stone heavy enough to require handling individual pieces by mechanical means, generally 1 ton and up.

STORM SURGE - A rise above normal water level on the open coast due to the action of wind stress on the water surface. Storm surge resulting from a hurricane also includes that rise in level due to atmospheric pressure reduction as well as that due to wind stress. See WIND SETUP.

STORM TIDE - See STORM SURGE.

A-36

STRAIT - A relatively narrow waterway between two larger bodies of water
See also SOUND.

STREAM - (1) A course of water flowing along a bed in the earth. (2) A
current in the sea formed by wind action, water density differences, etc. (Gulf Stream). See also CURRENT, STREAM.

SURF - The wave activity in the area between the shoreline and the
outermost limit of breakers.

SURF BEAT - Irregular oscillations of the nearshore water level, with
periods of the order of several minutes.

SURF ZONE - The area between the outermost breaker and the limit of
wave uprush. (See Figures A-2 and A-5.)

SURGE - (1) The name applied to wave motion with a period intermediate
between that of the ordinary wind wave and that of the tide, say
from 1/2 to 60 minutes. It is of low height; usually less than
0.3 foot. See also SEICHE. (2) In fluid flow, long interval
variations in velocity and pressure, not necessarily periodic,
perhaps even transient in nature. (3) See STORM SURGE.

SURGING BREAKER - See BREAKER.

SUSPENDED LOAD - (1) The material moving in suspension in a fluid, being
kept up by the upward components of the turbulent currents or by
colloidal suspension. (2) The material collected in or computed
from samples collected with a suspended load sampler. (A suspended load sampler is a sampler which attempts to secure a sample
of the water with its sediment load without separating the sediment from the water.) Where it is necessary to distinguish
between the two meanings given above, the first one may be called
the "true suspended load".

SWALE - The depression between two beach ridges.

SWASH - The rush of water up onto the beach face following the breaking
of a wave. Also UPRUSH, RUNUP. (See Figure A-2.)

SWASH CHANNEL - (1) On the open shore, a channel cut by flowing water in
its return to the parent body (e.g., a rip channel). (2) A secondary channel passing through or shoreward of an inlet or river
bar. (See Figure A-9.)

SWASH MARK - The thin wavy line of fine sand, mica scales, bits of sea-
weed, etc., left by the uprush when it recedes from its upward
limit of movement on the beach face.

SWELL - Wind-generated waves that have traveled out of their generating area. Swell characteristically exhibits a more regular and longer period, and has flatter crests than waves within their fetch (SEAS).

SYNOPTIC CHART - A chart showing the distribution of meterological conditions over a given area at a given time. Popularly called a weather map.

SYZYGY - The two points in the moon's orbit when the moon is in conjunction or opposition to the sun relative to the earth; time of new or full moon in the cycle of phases.

TERRACE - A horizontal or nearly horizontal natural or artificial topographic feature interrupting a steeper slope, sometimes occurring in a series.

THALWEG - In hydraulics, the line joining the deepest points of an inlet or stream channel.

TIDAL CURRENT - See CURRENT, TIDAL.

TIDAL DATUM - See CHART DATUM and DATUM PLANE.

TIDAL DAY - The time of the rotation of the earth with respect to the moon, or the interval between two successive upper transits of the moon over the meridian of a place, approximately 24.84 solar hours (24 hours and 50 minutes) or 1.035 times the mean solar day. (See Figure A-10.) Also called lunar day.

TIDAL FLATS - Marshy or muddy land areas which are covered and uncovered by the rise and fall of the tide.

TIDAL INLET - (1) A natural inlet maintained by tidal flow. (2) Loosely, any inlet in which the tide ebbs and flows. Also TIDAL OUTLET.

TIDAL PERIOD - The interval of time between two consecutive like phases of the tide. (See Figure A-10.)

TIDAL POOL - A pool of water remaining on a beach or reef after recession of the tide.

TIDAL PRISM - The total amount of water that flows into a harbor or estuary or out again with movement of the tide, excluding any freshwater flow.

TIDAL RANGE - The difference in height between consecutive high and low (or higher high and lower low) waters. (See Figure A-10.)

TIDAL RISE - The height of tide as referred to the datum of a chart. (See Figure A-10.)

TIDAL WAVE - (1) The wave motion of the tides. (2) In popular usage, any unusually high and destructive water level along a shore. It usually refers to STORM SURGE or TSUNAMI.

TIDE - The periodic rising and falling of the water that results from gravitational attraction of the moon and sun and other astronomical bodies acting upon the rotating earth. Although the accompanying horizontal movement of the water resulting from the same cause is also sometimes called the tide, it is preferable to designate the latter as TIDAL CURRENT, reserving the name TIDE for the vertical movement.

TIDE, DAILY RETARDATION OF - The amount of time by which corresponding tides grow later day by day (about 50 minutes).

TIDE, DIURNAL - A tide with one high water and one low water in a tidal day. (See Figure A-10)

TIDE, EBB - See EBB TIDE.

TIDE, FLOOD - See FLOOD TIDE.

TIDE, MIXED - See MIXED TIDE.

TIDE, NEAP - See NEAP TIDE.

TIDE, SEMIDIURNAL - See SEMIDIURNAL TIDE.

TIDE, SLACK - See SLACK TIDE.

TIDE, SPRING - See SPRING TIDE.

TIDE STATION - A place at which tide observations are being taken. It is called a *primary* tide station when continuous observations are to be taken over a number of years to obtain basic tidal data for the locality. A *secondary* tide station is one operated over a short period of time to obtain data for a specific purpose.

TIDE, STORM - See STORM SURGE.

TOMBOLO - A bar or spit that connects or "ties" an island to the mainland or to another island. (See Figure A-9.)

TOPOGRAPHY - The configuration of a surface, including its relief, the position of its streams, roads, building, etc.

TRAINING WALL - A wall or jetty to direct current flow.

TRANSITIONAL ZONE (TRANSITIONAL WATER) - In regard to progressive gravity waves, water whose depth is less than 1/2 but more than 1/25 the wavelength. Often called SHALLOW WATER.

TRANSLATORY WAVE - See WAVE OF TRANSLATION.

TRANSPOSED HURRICANE - See HYPOTHETICAL HURRICANE.

TROCHOIDAL WAVE - A theoretical, progressive oscillatory wave first proposed by Gerstner in 1802 to describe the surface profile and particle orbits of finite amplitude, nonsinusoidal waves. The wave form is that of a prolate cycloid or trochoid, and the fluid particle motion is rotational as opposed to the usual irrotational particle motion for waves generated by normal forces. See IRRO-TATIONAL WAVE

TROPICAL CYCLONE - See HURRICANE

TROPICAL STORM - A tropical cyclone with maximum winds less than 75 mph.

TROUGH OF WAVE - The lowest part of a wave form between successive crests. Also that part of a wave below stillwater level. (See Figure A-3.)

TSUNAMI - A long-period wave caused by an underwater disturbance such as a volcanic eruption or earthquake. Commonly miscalled "tidal wave".

TYPHOON - See HURRICANE.

UNDERTOW - A seaward current near the bottom on a sloping inshore zone. It is caused by the return, under the action of gravity, of the water carried up on the shore by waves. Often a misnomer for RIP CURRENT.

UNDERWATER GRADIENT - The slope of the sea bottom. See also SLOPE.

UNDULATION - A continuously propagated motion to and fro, in any fluid or elastic medium, with no permanent translation of the particles themselves.

UPCOAST - In United States usage, the coastal direction generally trending toward the north.

UPDRIFT - The direction opposite that of the predominant movement of littoral materials.

UPLIFT - The upward water pressure on the base of a structure or pavement.

UPRUSH - The rush of water up onto the beach following the breaking of a wave. Also SWASH, RUNUP. (See Figure A-2.)

VALLEY, SEA - A submarine depression of broad valley form without the steep side slopes which characterize a canyon.

VALLEY, SUBMARINE - A prolongation of a land valley into or across a continental or insular shelf, which generally gives evidence of having been formed by stream erosion.

VARIABILITY OF WAVES - (1) The variation of heights and periods between individual waves within a wave train. (Wave trains are not composed of waves of equal height and period, but rather of heights and periods which vary in a statistical manner.) (2) The variation in direction of propagation of waves leaving the generating area. (3) The variation in height along the crest, usually called "variation along the wave".

VELOCITY OF WAVES - The speed at which an individual wave advances. See WAVE CELERITY.

VISCOSITY - (or internal friction) - That molecular property of a fluid that enables it to support tangential stresses for a finite time and thus to resist deformation.

WATERLINE - A juncture of land and sea. This line migrates, changing with the tide or other fluctuation in the water level. Where waves are present on the beach, this line is also known as the limit of backrush. (Approximately the intersection of the land with the stillwater level.)

WAVE - A ridge, deformation, or undulation of the surface of a liquid.

WAVE AGE - The ratio of wave speed to wind speed.

WAVE, CAPILLARY - See CAPILLARY WAVE.

WAVE CELERITY - Wave speed.

WAVE CREST - See CREST OF WAVE.

WAVE CREST LENGTH - See CREST LENGTH, WAVE.

WAVE, CYCLOIDAL - See CYCLOIDAL WAVE.

WAVE DECAY - See DECAY OF WAVES.

WAVE DIRECTION - The direction from which a wave approaches.

WAVE FORECASTING - The theoretical determination of future wave characteristics, usually from observed or predicted meteorological phenomena.

WAVE GENERATION - See GENERATION OF WAVES.

WAVE, GRAVITY - See GRAVITY WAVE.

WAVE GROUP - A series of waves in which the wave direction, wavelength, and wave height vary only slightly. See also GROUP VELOCITY.

WAVE HEIGHT - The vertical distance between a crest and the preceding trough. See also SIGNIFICANT WAVE HEIGHT. (See Figure A-3.)

WAVE HEIGHT COEFFICIENT - The ratio of the wave height at a selected point to the deepwater wave height. The refraction coefficient multiplied by the shoaling factor.

WAVE HINDCASTING - See HINDCASTING, WAVE.

WAVE, IRROTATIONAL - See IRROTATIONAL WAVE.

WAVELENGTH - The horizontal distance between similar points on two successive waves measured perpendicular to the crest. (See Figure A-3.)

WAVE, MONOCHROMATIC - See MONOCHROMATIC WAVE.

WAVE, OSCILLATORY - See OSCILLATORY WAVE.

WAVE PERIOD - The time for a wave crest to traverse a distance equal to one wavelength. The time for two successive wave crests to pass a fixed point. See also SIGNIFICANT WAVE PERIOD.

WAVE, PROGRESSIVE - See PROGRESSIVE WAVE.

WAVE PROPAGATION - The transmission of waves through water.

WAVE RAY - See ORTHOGONAL.

WAVE, REFLECTED - That part of an incident wave that is returned seaward when a wave impinges on a steep beach, barrier, or other re flecting surface.

WAVE REFRACTION - See REFRACTION OF WATER WAVES.

WAVE SETUP - See SETUP, WAVE.

WAVE, SINUSOIDAL - An oscillatory wave having the form of a sinusoid.

WAVE, SOLITARY - See SOLITARY WAVE

WAVE SPECTRUM - In ocean wave studies, a graph, table, or mathematical equation showing the distribution of wave energy as a function of wave frequency. The spectrum may be based on observations or theoretical considerations. Several forms of graphical display are widely used.

WAVE, STANDING - See STANDING WAVE.

WAVE STEEPNESS - The ratio of the wave height to the wavelength.

WAVE TRAIN - A series of waves from the same direction.

WAVE OF TRANSLATION - A wave in which the water particles are permanently displaced to a significant degree in the direction of wave travel. Distinguished from an OSCILLATORY WAVE.

WAVE, TROCHOIDAL - See TROCHOIDAL WAVE.

WAVE TROUGH - The lowest part of a wave form between successive crests. Also that part of a wave below stillwater level.

WAVE VARIABILITY - See VARIABILITY OF WAVES.

WAVE VELOCITY - The speed at which an individual wave advances.

WAVE, WIND - See WIND WAVES.

WAVES, INTERNAL - See INTERNAL WAVES.

WEIR JETTY - An updrift jetty with a low section or weir over which littoral drift moves into a predredged deposition basin which is dredged periodically.

WHARF - A structure built on the shore of a harbor, river, or canal, so that vessels may lie alongside to receive and discharge cargo and passengers.

WHITECAP - On the crest of a wave, the white froth caused by wind.

WIND CHOP - See CHOP.

WIND, FOLLOWING - See FOLLOWING WIND.

WIND, OFFSHORE - A wind blowing seaward from the land in a coastal area.

WIND, ONSHORE - A wind blowing landward from the sea in a coastal area.

WIND, OPPOSING - See OPPOSING WIND.

WIND SETUP - (1) The vertical rise in the stillwater level on the leeward side of a body of water caused by wind stresses on the surface of the water. (2) The difference in stillwater levels on the windward and the leeward sides of a body of water caused by wind stresses on the surface of the water. (3) Synonymous with WIND TIDE and STORM SURGE. STORM SURGE is usually reserved for use on the ocean and large bodies of water. WIND SETUP is usually reserved for use on reservoirs and smaller bodies of water. (See Figure A-11.)

WIND TIDE - See WIND SETUP, STORM SURGE.

WINDWARD - The direction from which the wind is blowing.

WIND WAVES - (1) Waves being formed and built up by the wind. (2) Loosely any wave generated by wind.

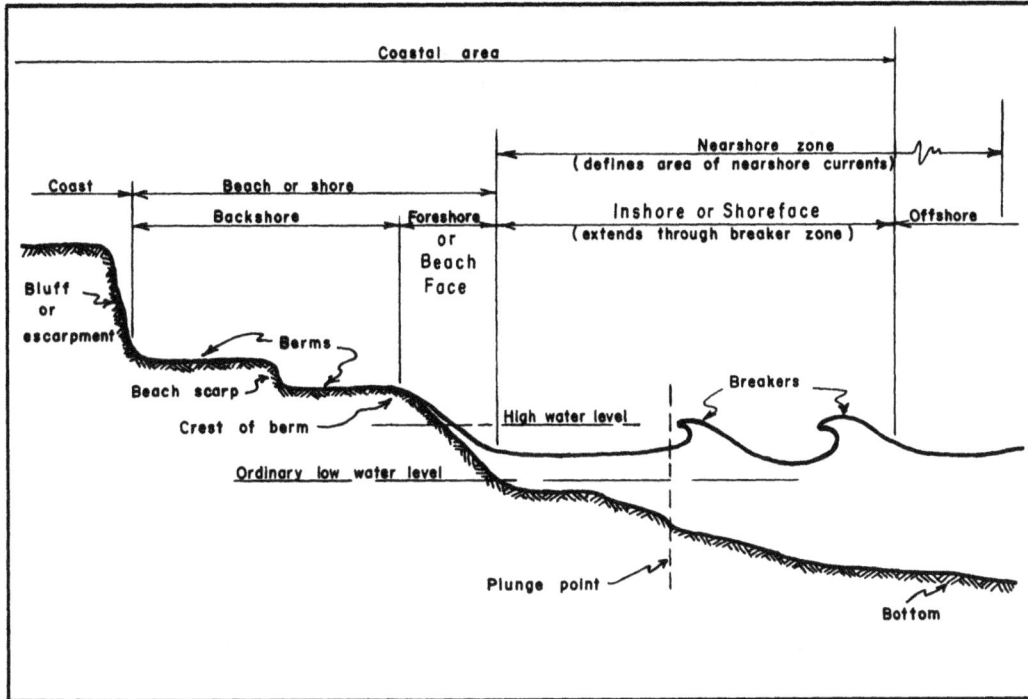

Figure A-1. Beach Profile-Related Terms.

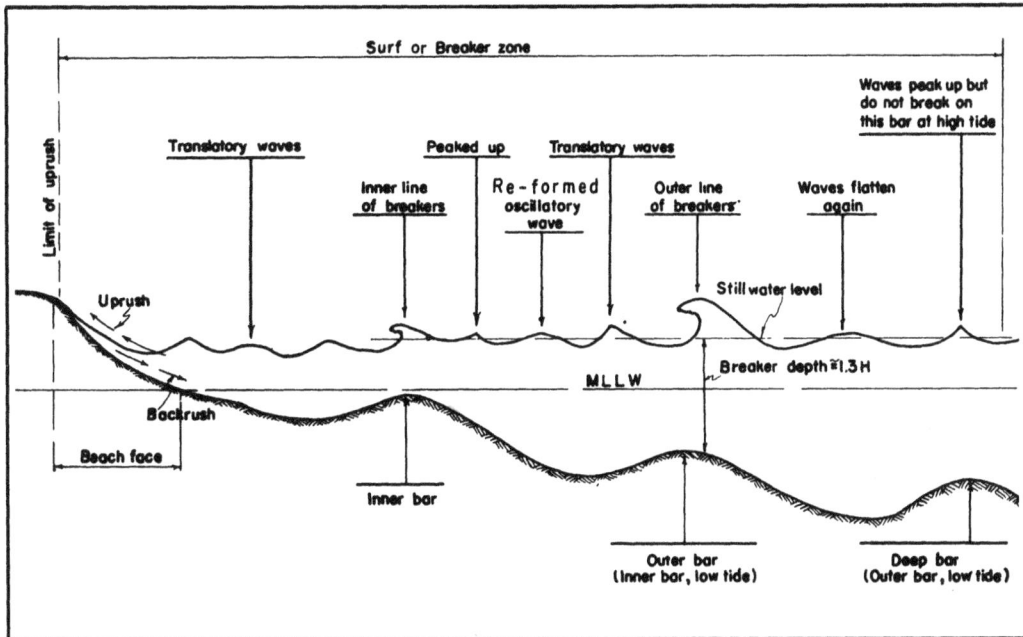

Figure A-2. Schematic Diagram of Waves in the Breaker Zone.

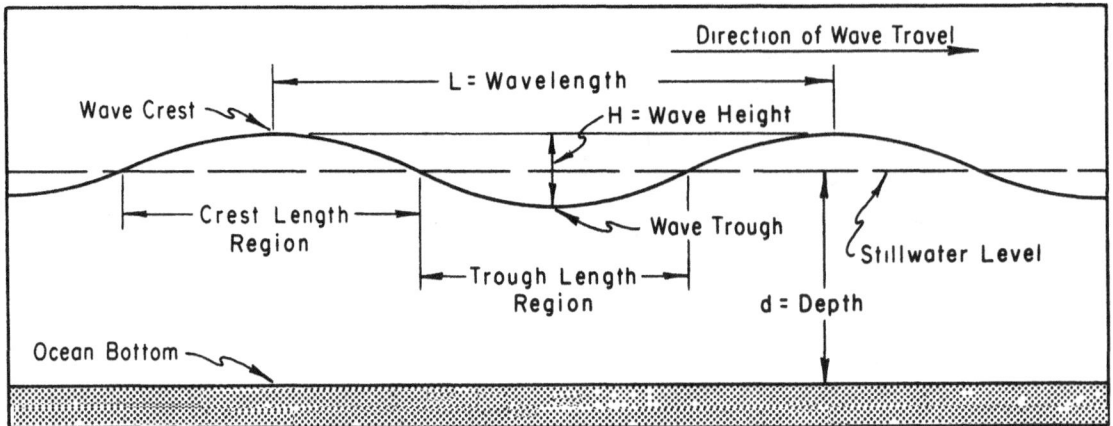

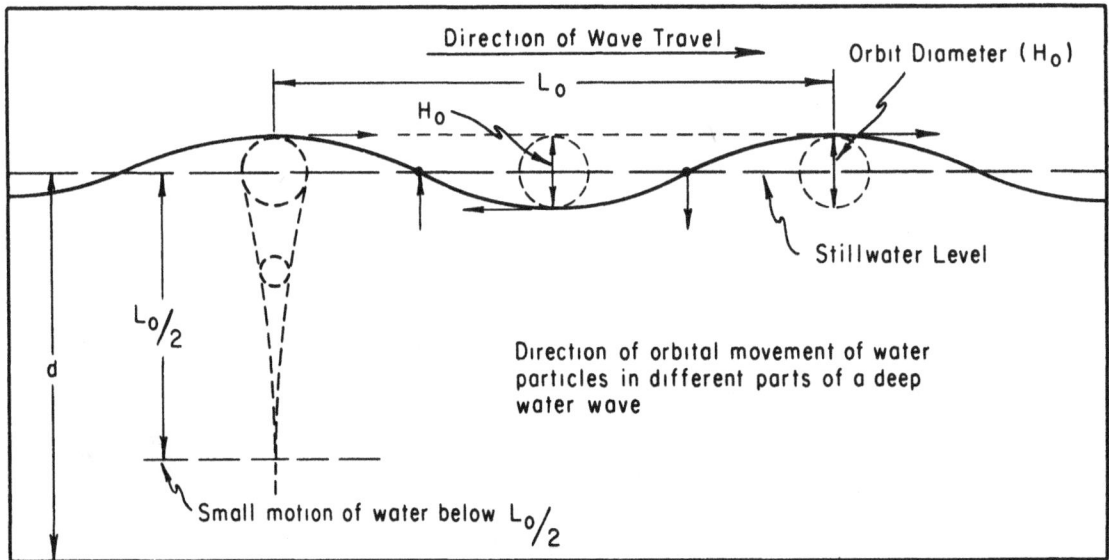

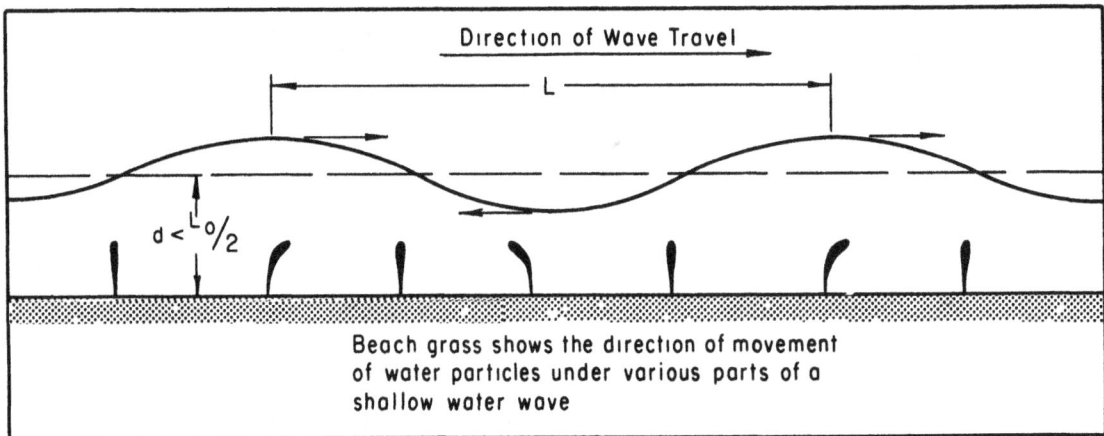

Figure A-3. Wave Characteristics and Direction of
Water Particle Movement.

SPILLING BREAKER

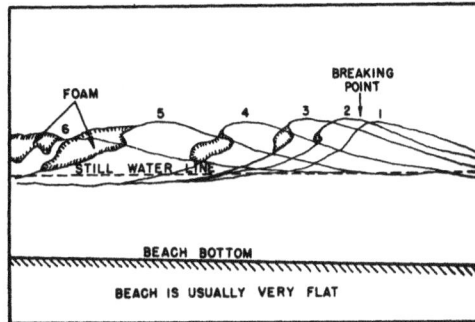

SKETCH SHOWING THE GENERAL CHARACTER
OF SPILLING BREAKERS

PLUNGING BREAKER

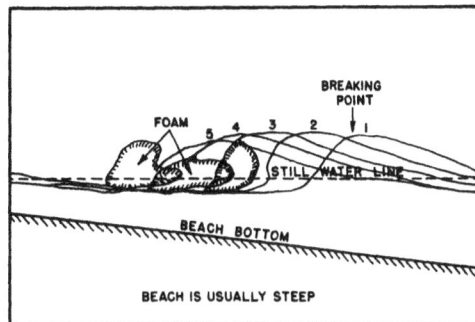

SKETCH SHOWING THE GENERAL CHARACTER
OF PLUNGING BREAKERS

SURGING BREAKER

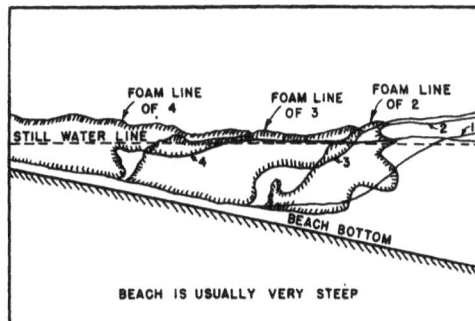

SKETCH SHOWING THE GENERAL CHARACTER
OF SURGING BREAKERS

Both photographs and diagrams of the three types of breakers are
presented above. The sketches consist of a series of profiles of the
wave form as it appears before breaking, during breaking and after
breaking. The numbers opposite the profile lines indicate the relative
times of occurences.

(Wiegel, 1953)

Figure A-4.   Breaker Types.

WIDE
SURF ZONE

NARROW
SURF ZONE

HIGH WAVES
ON POINT

SUBMARINE
RIDGE

NARROW
SURF ZONE

WIDE
SURF ZONE

WIDE
SURF ZONE

POINT

NARROW
SURF ZONE

Halfmoon Bay, California

Note the increasing width of the surf zone with increasing degree
of exposure to the south

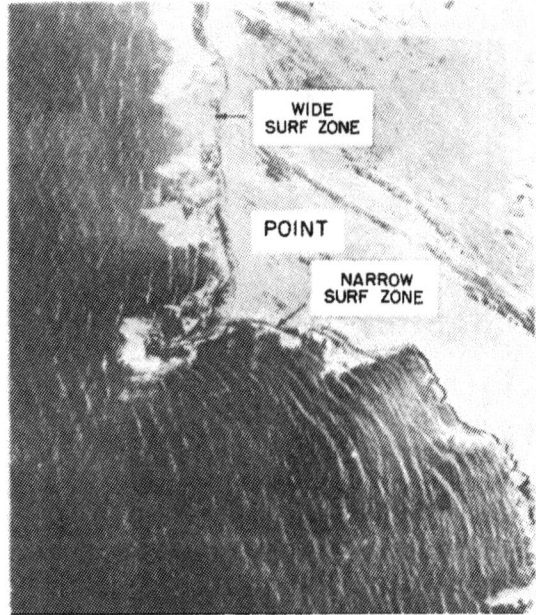

Purisima Pt, California

Refraction of waves around a headland produces low waves and
a narrow surf zone where bending is greatest

(Wiegel,1953)

Figure A-5. Refraction of Waves.

A-48

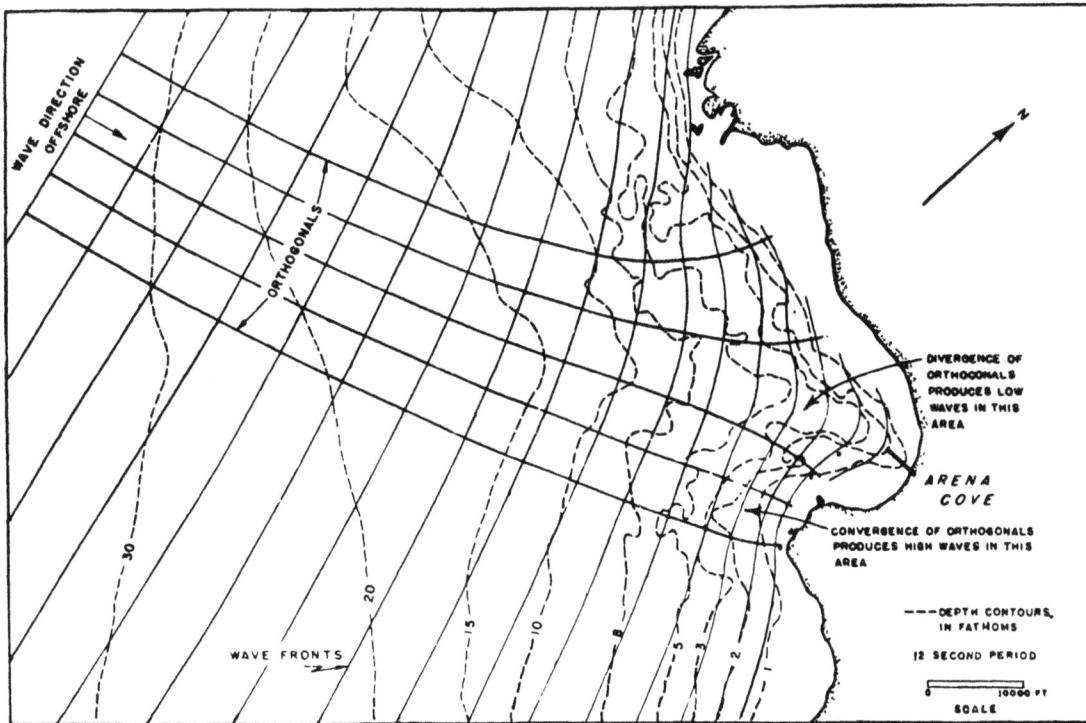

Figure A-6. Refraction Diagram.

(Wiegel, 1953)

A-49

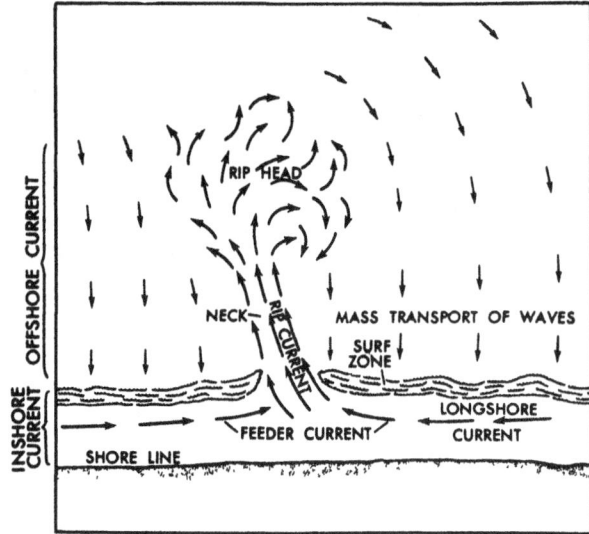

Nearshore Current System
USNOO   SP 35

after Wiegel 1953

Figure A-7.   Beach Features.

Figure A-8.   Shoreline Features.

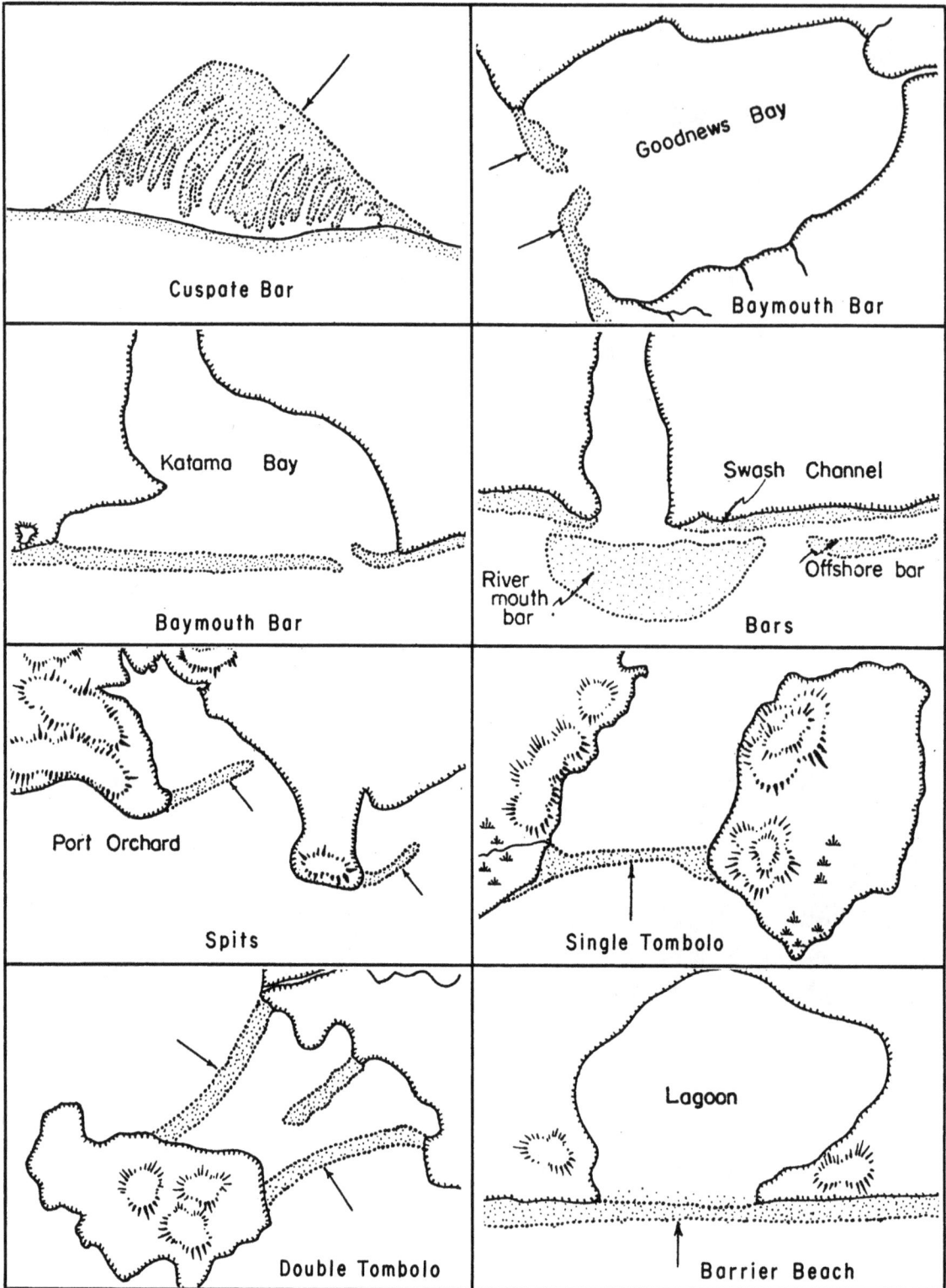

Figure A-9.   Bar and Beach Forms.   (from D. W. Johnson, Shore Process and Shoreline Development, 1919, published by John Wiley & Sons, Inc.)

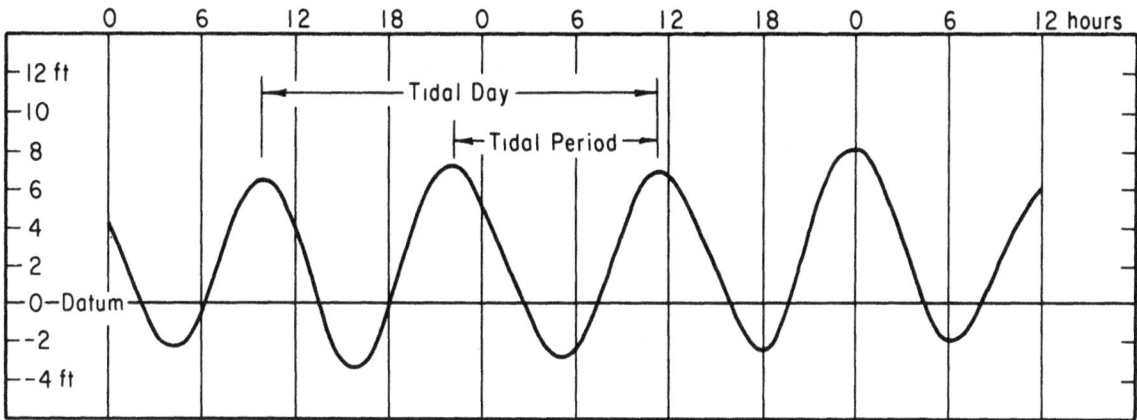

SEMIDIURNAL

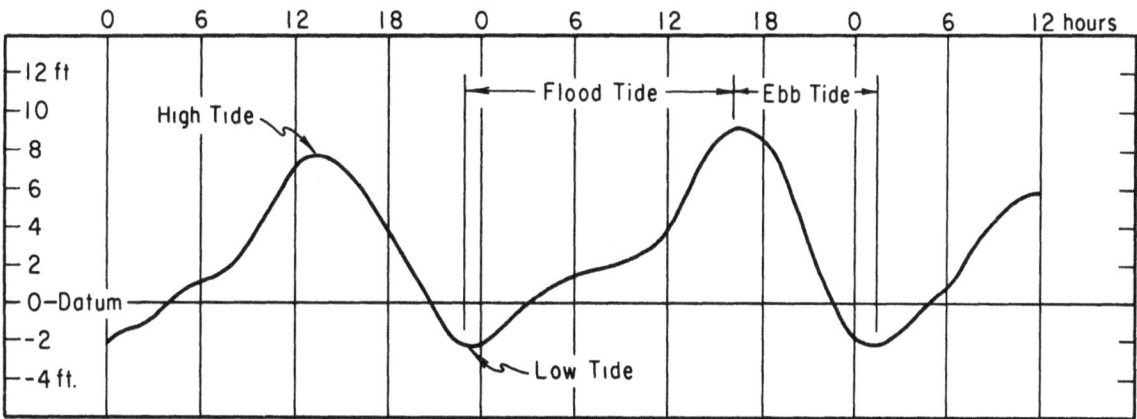

DIURNAL

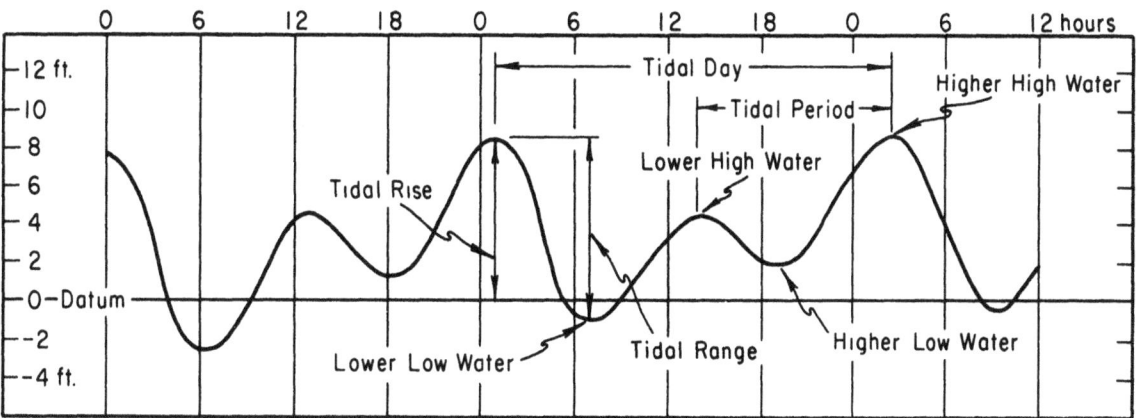

MIXED

(Wiegel, 1953)

Figure A-10. Types of Tides.

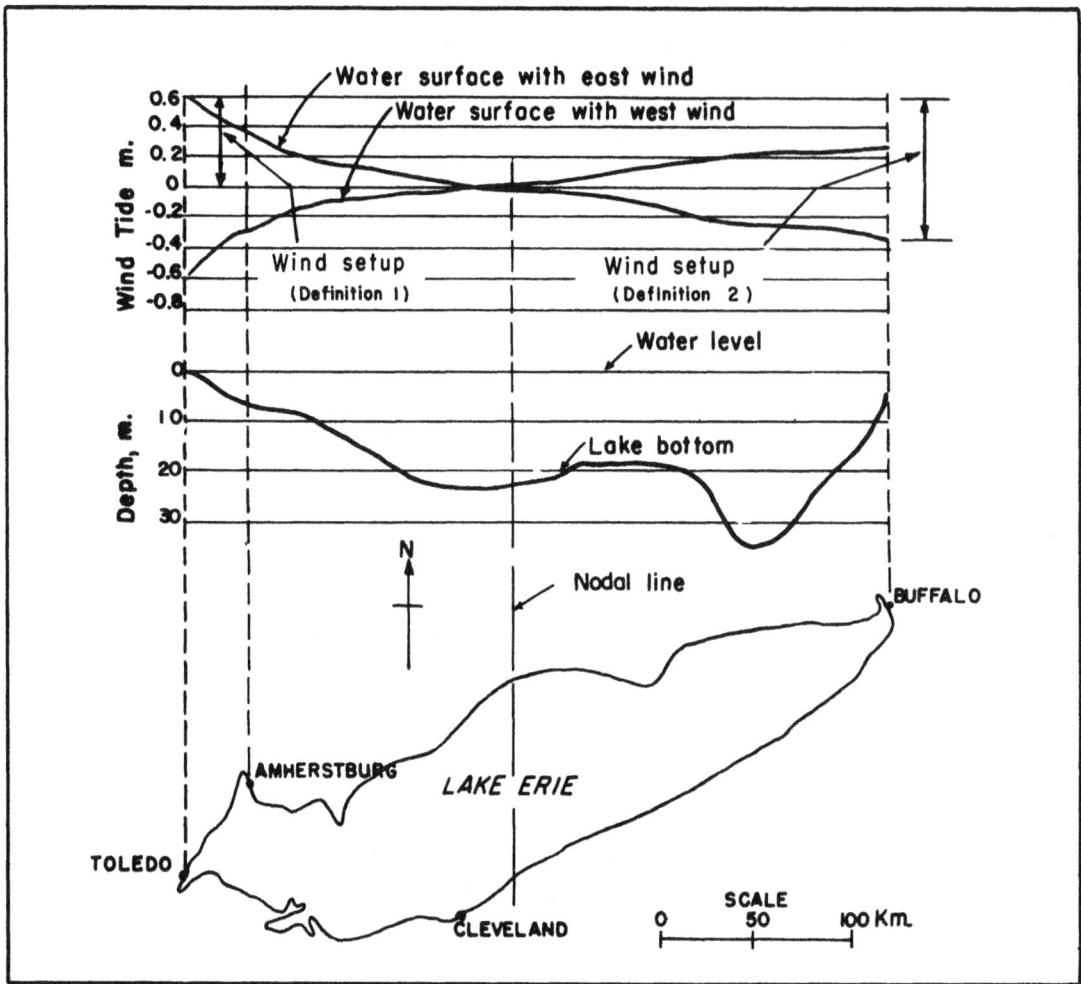

Figure A-11.   Setup or Wind Tide.

APPENDIX B

LIST

OF

SYMBOLS

APPENDIX B

LIST OF SYMBOLS

Symbol	Definition	Dimension	Example Unit
A	Friction loss parameter (eq. 3-41)	——	—————
	●Area	L^2	$ft.^2$, $mi.^2$
	●Constant = 7500 (eq. 4-41)	L^3/F	(yd.^{3}-sec.)/(lbs.-yr.)
	●Constant = 0.0161 (eq. 3-23)	——	—————
	●Kinematic wind stress (eq. 3-68)	$(L/T)^2$	$(mi./hr.)^2$, $(knots)^2$
	●Major ellipse semiaxis of wave particle movement (eq. 2-22)	L	feet
a	Wave form amplitude	L	feet
	●Breaking wave dynamic moment reduction factor for low wall	——	—————
	●Breaker height parameter (eq. 2-92)	——	—————
a_j	Amplitude of jth wave in series	L	feet
B	Breakwater gap width	L	feet
	●Minor ellipse semiaxis of wave particle movement (eq. 2-23)	L	feet
	●Kinematic wind stress (eq. 3-69)	$(L/T)^2$	$(mi./hr.)^2$, $(knots)^2$
	●Constant such as 0.3692 (eq. 3-23)	——	—————
	●Rubble structure crest width	L	feet
	●Rubble crest width in front of wall	L	feet
B'	Effective breakwater gap width	L	feet
b	Spacing between wave orthogonals	L	feet
	●Breaker height parameter (eq. 2-93)	——	—————
	●Width of enclosed basin (eq. 3-84)	L	miles
	●Structure crest width (eq. 7-9)	L	feet
	●Height of overtopped wall, seafloor to wall crest (eq. 7-70)	L	feet
	●Height of rubble base alone (eq. 7-75)	L	feet
b'	Overtopped wall height above wave trough	L	feet
b_i	Length of shoreline considered as line source for littoral zone sediment	L	feet
b_o	Orthogonal spacing, deep water	L	feet
b_1	Orthogonal spacing at a diffracted wave crest	L	feet
b_2	Orthogonal spacing of a progressing wave undergoing diffraction at a second point shoreward	L	feet

Symbol	Definition	Dimension	Example Unit
C	Wave velocity; phase velocity	L/T	ft./sec.
	●Constant such as 2.2024 (eq. 3-23)	——	————
	●Coefficient such as 1.165×10^{-3} (eq. 3-83)	——	————
C_D	Drag Coefficient	——	————
C_g	Group velocity	L/T	ft./sec.
C_L	Lift coefficient	——	————
C_M	Mass or inertia coefficient	——	————
C_o	Deepwater wave velocity	L/T	ft./sec.
cn	Jacobian elliptical cosine function	——	————
D	Total water depth, includes surge	L	feet
	●Depth one wavelength in front of wall (eq. 7-76)	L	feet
	●Duration of an observation	T	sec.; hr.
	●Decay distance	L	nautical mile
	●Constant such as 0.8798 (eq. 3-23)	——	————
	●Pile diameter	L	feet
	●Percent damage to rubble structure (Table 7-7)	——	————
d	Water depth (bed to SWL)	L	feet
	●Grain diameter	L	millimeters
d_b	Depth of water at breaking wave	L	feet
d_g	Equivalent stone diameter (eq. 7-115)	L	feet
d_s	Water depth at toe of structure	L	feet
$\bar{d}_T$	Average total depth—includes astronomical tide and surge (Table 3-4)	L	feet
d_x	Water depth (MLW) at specified location (Table 3-4)	L	feet
d_1	Depth below SWL of rubble foundation crest (Figure 7-99)	L	feet
d_1, d_2	Depths at specified points (including astronomical tide and surge) (Table 3-4)	L	feet
d_{50}	Size in millimeters of 50th percentile of sediment sample ($d_{50} = M_d$)	L	millimeters

Symbol	Definition	Dimension	Example Unit
E	Total energy in one wavelength per unit crest width	LF/L	ft.-lbs./ft. crest width
	●Crest elevation of structure above MLW or other datum plane	L	feet
$\bar{E}$	Total average wave energy per unit surface area; specific energy; energy density	LF/L^2	ft.-lbs./ft.^{2}
$(\bar{E})_A$	Average wave energy per unit water surface area for several waves	LF/L^2	ft.-lbs./ft.^{2}
E_a	Longshore component of wave energy (eq. 4-39)	LF/LT	ft.-lbs./ft./day
E_k	Kinetic energy in one wavelength per unit crest width	LF/L	ft.-lbs./ft. crest width
E(k)	Complete elliptic integral of second kind	——	————
E_o	Deepwater wave energy	LF/L	ft.-lbs./ft. crest width
E_p	Potential energy in one wavelength per unit crest width	LF/L	ft.-lbs./ft. crest width
F	Fetch length	L	nautical mile
	●Total horizontal force acting about mudline on pile at given instant	F	pounds
	●Nonbreaking, nonovertopping wave force on wall extending full water depth	F/L	lbs./ft. of wall
F'	(Reduced) Force on overtopped wall which extends full water depth (eq. 7-69)	F/L	lbs./ft. of wall
F''	(Reduced) Force on wall resting on rubble foundation (eq. 7-73)	F/L	lbs./ft. of wall
F_c	Total horizontal force per unit length of wall from nonbreaking wave crest	F/L	lbs./ft. of wall
F_D	Total horizontal drag force on a pile at given instant	F	pounds
F_{Dm}	Maximum value of F_D for given wave	F	pounds
F_E	Effective fetch length due to limited width	L	miles
F_e	Effective fetch length, unrestricted body of water	L	nautical mile
F_e'	Equivalent effective fetch length (Table 3-4)	L	nautical mile
F_i	Total horizontal inertial force on pile at given instant	F	pounds

Symbol	Definition	Dimension	Example Unit
F_{im}	Maximum value of F_i for given wave	F	pounds
F_L	Lift force (lateral force on pile from flow velocity)	F	pounds
F_{Lm}	Maximum lift force for given wave	F	pounds
F_M	Surge correction factor for storm speed, angle to coast (eq. 3-78)	---	-----
F_m	Maximum total force on pile	F	pounds
F_m or F_{min}	Minimum fetch length	L	nautical mile
F_o	Dimensionless fall time parameter (eq. 4-20)	---	-----
F_s	Surge correction factor for shoaling effects (eq. 3-78)	---	-----
F_t	Total horizontal force per unit length of wall subjected to non-breaking wave trough	F/L	lbs./ft. of wall
F_{Total}	Total force on pile group	F	pounds
f	Coriolis parameter	T^{-1}	seconds^{-1}, hours^{-1}
	●Wave frequency	T^{-1}	seconds^{-1}
	●Horizontal force per unit length of pile	F/L	lb./ft.
	●Decimal frequency (eq. 4-41)	---	-----
	●Weight factor of armor unit (Figure 7-87)	---	-----
f_D	Horizontal drag force per unit length of vertical pile	F/L	lbs./ft.
f_{Dm}	Maximum value of f_D	F/L	lbs./ft.
f_f	Bottom friction factor; Darcy-Weisbach friction factor (eq. 3-41; eq. 3-57)	---	-----
f_i	Horizontal inertial force per unit length of vertical pile	F/L	lbs./ft.
f_{im}	Maximum value of f_i	F/L	lbs./ft.
f_m	Maximum force per unit length of pile	F/L	lbs./ft.
G	Coefficient (eq. 3-91)	---	-----
g	Gravitational acceleration	L/T^2	ft./sec.^{2}
~g	Subscript for: Group	---	-----
	● Gage	---	-----
	● Gross	---	-----

Symbol	Definition	Dimension	Example Unit
H	Wave Height	L	feet
	• Design wave height—wave height for which structure is designed; maximum wave height causing no damage or damage within specified limits	L	feet
	• High pressure area on weather maps	F/L^2	millibars, inches of mercury
$\bar{H}$	Average wave height; $\bar{H} = 0.886\ H_{rms}$	L	feet
$\hat{H}$	Arbitrary wave height for probability distributions	L	feet
H_b	Wave height at breaking (breaker height)	L	feet
H_D	Significant wave height, end of decay distance	L	feet
H_F	Significant wave height at downwind end of fetch	L	feet
H_g	Gage wave height	L	feet
H_i	Incident wave height	L	feet
H_j	Height of jth wave in series	L	feet
H_{max}	Maximum wave height for specified period of time	L	feet
H_n	Most probable n^{th} highest wave height	L	feet
H_o	Deepwater significant wave height	L	feet
H_o'	Deepwater wave height equivalent to observed shallow water wave if unaffected by refraction and friction; $H_o' = H_o K_f K_R = H/K_s$	L	feet
H_r	Reflected wave height	L	feet
H_{rms}	Root mean square wave height	L	feet
H_s	Significant wave height; $H_{1/3}$; average height of highest one-third of waves for specified time period	L	feet
$\bar{H}_s$	Mean significant wave height (eq. 4-7)	L	feet
$\hat{H}_s$	Arbitrary significant wave height for probability distributions (eq. 4-6)	L	feet
$H_{s\ min}$	Approximate minimum significant wave height from a distribution of significant heights (eq. 4-6)	L	feet

Symbol	Definition	Dimension	Example Unit
H_t	Wave height transmitted past obstacle	L	feet
H_w	Wave height at wall	L	feet
$H_{1/3}$	Significant wave height; H_s	L	feet
H_1	Average height of highest 1 percent of all waves for given time period	L	feet
H_{10}	Average height of highest 10 percent of all waves for given time period	L	feet
h	Range of tide	L	feet
	● Height of retaining wall	L	feet
	● Height of backfill at wall if lower than wall	L	feet
	● Structure height, toe to crest	L	feet
	● Vertical distance from dune base or berm crest to depth of seaward limit of significant longshore transport (Figure 4-44)	L	feet
h'	Broken wave height above ground surface at structure toe landward of SWL	L	feet
h_c	Height of broken wave above SWL	L	feet
h_o	Height of clapotis orbit center above SWL	L	feet
I_ℓ	Submerged weight of longshore transport	F/T	lbs./yr.
i	Angle of backfill surface from horizontal	---	degrees
$\sim i$	Subscript for discrete points in space (eq. 3-65)	---	-----
	● Subscript for an incident wave characteristic	---	-----
	● Subscript dummy variable for use in sediment budget analysis	---	-----
J	Distance between bottom contours (Figure 2-21)	L	feet
j	Dummy variable	---	-----
$\sim j$	Subscript dummy variable	---	-----
K	Pressure response factor at bottom (eq. 2-31)	---	-----
	● Bottom friction coefficient (eq. 3-57)	---	-----
	● Constant such as 6.5882 (eq. 3-23)	---	-----
	● Constant for Rankine vortex model of hurricane wind field (eq. 3-27)	T^{-1}	second^{-1}
	● Coefficient dependent on breaker height-to-depth ratio and ratio of trough depression to breaker height (eq. 4-18)	---	-----

Symbol	Definition	Dimension	Example Unit
K	Diffraction coefficient	--	-----
K_D	Armor unit stability coefficient (eq. 7-105)	--	-----
	●Dimensionless factor for calculation of total drag force on pile at given phase (eq. 7-23)	--	-----
K_{Dm}	Maximum value of K_D for given wave	--	-----
K_f	Wave height reduction factor from friction; friction factor (Figure 3-35, Table 3-4)	--	-----
K_i	Dimensionless factor for calculation of total inertial force on pile at given phase (eq. 7-22)	--	-----
K_{im}	Maximum value of K_i for given wave	--	-----
$K_{(k)}$	Complete elliptic integral of the first kind	--	-----
K_R	Refraction coefficient	--	-----
K_{RR}	Stability coefficient for smooth relatively rounded graded riprap armor units (eq. 7-106)	--	-----
K_s	Shoaling coefficient (eq. 2-44)	--	-----
$(K_s)_b$	Shoaling coefficient at breaker position	--	-----
$(K_s)_g$	Shoaling coefficient at gage recorder	--	-----
K_{s2}	Revised shoaling coefficient after friction effects over continental shelf (Table 3-4)	--	-----
K_z	Pressure response factor at any depth z (eq. 2-29)	--	-----
K_1, K_2	Constants of wind-surface friction (eq. 3-59)	--	-----
k	Wave number ($2\pi/L$)	L^{-1}	$feet^{-1}$
	●Runup correction factor for scale effects	--	-----
	●Wind stress coefficient (surface friction coefficient) (eq. 3-59)	--	-----
	●Modulus of elliptic integrals	--	-----
	●Kip: 1000 pounds	F	kips
	●Proportionality constant between longshore wave energy P_s and submerged weight transport I	--	-----
k	Wind correction factor for overtopping rates (eq. 7-8)	--	-----
	●Constant such as 0.003 (eq. 3-82)	--	-----

Symbol	Definition	Dimension	Example Unit
k_i	Source (or sink) fraction of gross longshore transport rate (eq. 4-47)	--	----
k_Δ	Layer coefficient of rubble structure	--	----
L	Wavelength ●Low pressure on weather map	L F/L^2	feet millibars, inches of mercury
L_A	Wavelength in given depth according to linear (Airy) theory; L_A may differ from L (eq. 7-14)	L	feet
L_b	Wavelength at breaking	L	feet
L_c	Width of caisson	L	feet
L_D	Wavelength in water depth D (eq. 7-76)	L	feet
L_d	Wavelength in water depth d_s (eq. 7-79)	L	feet
L_o	Deepwater wavelength	L	feet
ℓ_B	Enclosed basin length (eq. 3-42)	L	feet
$\ell_{B'}$	Length of rectangular basin open at one end (eq. 3-44)	L	feet
ℓ_n	Distance from reference pile to n^{th} pile of pile group (eq. 7-49)	L	feet
$\sim\ell_t$	Subscript for longshore transport to left as viewed from beach	--	----
M	Total wave moment about mudline on pile (eq. 7-21) ●Nonbreaking wave moment about toe of wall extending full depth of water ●Variable of solitary wave theory, function of H/d (eq. 2-67) ●Mean diameter of sediment sample	LF LF/L -- L	ft.-lbs. ft.-lbs./ft. of wall ---- millimeters
M'	Moment about toe of wall overtopped by nonbreaking wave (eq. 7-71)	LF/L	ft.-lbs./ft. of wall
M_A''	Moment about bottom (mudline) for wall on rubble foundation (eq. 7-74)	LF/L	ft.-lbs./ft. of wall
M_B''	Moment about base of wall on rubble foundation (eq. 7-75)	LF/L	ft.-lbs./ft. of wall
M_c	Total moment about toe of wall per unit length from nonbreaking wave crest	LF/L	ft.-lbs./ft. of wall

Symbol	Definition	Dimension	Example Unit
M_D	Total drag moment acting on pile about mudline (eq. 7-25)	LF	ft.-lbs.
M_{Dm}	Maximum value of M_D	LF	ft.-lbs.
M_d	Median diameter of sediment sample	L	millimeters
$M_{d\phi}$	Median diameter of sediment sample in phi units	L	phi
M_i	Total inertial moment acting on pile about mudline (eq. 7-24)	LF	ft.-lbs.
M_{im}	Maximum value of M_i for given wave	LF	ft.-lbs.
M_m	Maximum total moment on pile about mudline (eq. 7-36)	LF	ft.-lbs.
	●Maximum overturning moment about toe of wall from dynamic component of wave pressure (breaking or broken waves) (eq. 7-78)	LF/L	ft.-lbs./ft. of wall
M'_m	Reduced maximum moment against wall from breaking wave of height greater than wall (eq. 7-83)	LF/L	ft.-lbs./ft. of wall
M_s	Hydrostatic moment against wall from breaking or broken waves	LF/L	ft.-lbs./ft. of wall
M_t	Total moment about toe of wall per unit length from nonbreaking wave trough (Section 7.323)	LF/L	ft.-lbs./ft. of wall
	●Total moment about toe of wall per unit length from breaking or broken wave crest (eq. 7-81)	LF/L	ft.-lbs./ft. of wall
M_{Total}	Total moment on pile group about mudline (eq. 7-52)	LF	ft.-lbs.
M_{xx}	Momentum transport quantity per unit width (eq. 3-50)	L^2/T^2	ft.^{2}/sec.^{2}
M_{xy}	Momentum transport quantity per unit width (eq. 3-50)	L^2/T^2	ft.^{2}/sec.^{2}
M_{yy}	Momentum transport quantity per unit width (eq. 3-50)	L^2/T^2	ft.^{2}/sec.^{2}
M_ϕ	Mean diameter of sediment sample in phi units	L	phi
$M_{\phi b}$	Mean diameter (phi units) of borrow material (eq. 6-1)	L	phi

Symbol	Definition	Dimension	Example Unit
$M_{\phi n}$	Mean diameter (phi units) of native (beach) material (eq. 6-1)	L	phi
M_1	Coefficient determined by eq. 4-15	—	————
m	Position of wave in front of wave generator (Section 2.238) ● Beach slope	L L/L	wave lengths ft.(rise)/ft.(run)
N	Correction factor in determination of η (eta) from subsurface pressure (eq. 2-32) ● Variable in solitary wave theory (eq. 2-67) ● A total number of items	— — —	———— ———— ————
N_r	Required number of individual armor units (eq. 7-109)	—	————
N_s	Design stability number for rubble foundations and toe protection (eq. 7-110)	—	————
n	Number of layers of armor units in rubble structure protective cover ● Number of armor units across rubble structure crest ● Ratio of group velocity to individual wave velocity ● Number of seiche nodes along closed rectangular basin axis ● Degrees latitude (isobar spacing—not location) ● Dimensionless value $1 + (\tau_s/\tau_b)$; (eq. 3-82) ● A number ● Dummy variable	— — — — — — — —	———— ———— ———— ———— degrees ———— ———— ————
n′	Number of seiche nodes along rectangular basin open at one end excluding node at opening	—	————
$\sim_n$	Subscript referencing a particular pile in a pile group ● Subscript for net longshore transport rate ● Subscripted dummy variable	— — —	———— ———— ————
$\sim^n$	Superscript for discrete points in time (eq. 3-65)	—	————
n_o	Deepwater ratio of group velocity to individual wave velocity	—	————
$\sim_o$	Subscript for deepwater condition	—	————

Symbol	Definition	Dimension	Example Unit
P	Average porosity of rubble structure cover layer (eq. 7-109)	—	percent (%)
	●Precipitation rate (eq. 3-50)	L/T	in./hr.
$\bar{P}$	Wave power; average energy flux transmitted across a plane perpendicular to wave advance	LF/T/L	ft.-lbs./sec./ft. wave crest
P_a	Active earth force	F/L	lbs./ft. of wall
P_ℓ	Longshore component of wave energy flux (eq. 4-27)	LF/T/L	ft.-lbs./sec./ft. of beach
$P_{\ell s}$	Surf zone approximation of P_ℓ (eq. 4-28)	LF/T/L	ft.-lbs./sec./ft. of beach
$\bar{P}_o$	Deepwater $\bar{P}$	LF/T/L	ft.-lbs./sec./ft. wave crest
P_p	Passive earth force	F/L	lbs./ft. of wall
p	Gage pressure; pressure at any distance below fluid surface relative to surface	F/L^2	lbs./ft.^{2}
	●Atmospheric pressure at point located distance r from (hurricane) storm center (eq. 3-28)	F/L^2	inches of mercury; millibars
p'	Total or absolute subsurface pressure: includes dynamic, static and atmospheric pressures (eq. 2-26)	F/L^2	lbs./ft.^{2}
p_a	Atmospheric pressure (eq. 2-26)	F/L^2	lbs./ft.^{2}
p_m	Maximum dynamic pressure by breaking and broken waves on vertical wall (eq. 7-76)	F/L^2	lbs./ft.^{2}
p_n	Pressure at outskirts or periphery of storm	F/L^2	inches of mercury
	●Normal sea level atmospheric pressure = 29.92 inches Hg.(eq. 3-35)	F/L^2	inches of mercury
p_o	Central pressure of storm; CPI	F/L^2	inches of mercury
p_s	Maximum broken wave hydrostatic pressure against wall (eq. 7-89)	F/L^2	lbs./ft.^{2}
p_1	Nonbreaking wave pressure difference from stillwater hydrostatic pressure as clapotis crest (trough) passes (eq. 7-68)	F/L^2	lbs./ft.^{2}
Q	Overtopping rate	$L^3/T/L$	ft.^{3}/sec./ft. of wall
	●Volumetric flow rate for setup in long, closed basin	L^3/T	mi.^{3}/hr.
	●Rate at which littoral drift is moved parallel to shoreline; longshore transport rate	L^3/T	yd.^{3}/yr.

Symbol	Definition	Dimension	Example Unit
Q_c	Overtopping rate corrected for wind effects	$L^3/T/L$	ft.^{3}/sec./ft. of wall
Q_g	Gross longshore transport rate	L^3/T	yd.^{3}/yr.
Q_i^+	Point source for littoral zone sediment budget	L^3/T	yd.^{3}/yr.
Q_i^-	Point sink for littoral zone sediment budget	L^3/T	yd.^{3}/yr.
Q_i^{*+}	Line source total contribution to littoral zone sediment budget	L^3/T	yd.^{3}/yr.
Q_i^{*-}	Line sink total deduction from littoral zone sediment budget	L^3/T	yd.^{3}/yr.
Q_ℓ	Longshore transport rate ($Q_\ell = Q$)	L^3/T	yd.^{3}/yr.
Q_n	Net longshore transport rate	L^3/T	yd.^{3}/yr.
Q_o^*	Empirically determined overtopping coefficient (eq. 7-6)	--	----
q_i^+	Line source per unit length in littoral zone sediment budget	$L^3/T/L$	yd.^{3}/yr./ft. of beach
q_i^-	Line sink per unit length in littoral zone sediment budget	$L^3/T/L$	yd.^{3}/yr./ft. of beach
R	Wave runup	L	feet
	●Dynamic component of breaking or broken wave force per unit length of wall if wall is perpendicular to direction of wave advance (eq. 7-103)	F/L	lbs./ft. of wall
	●Resultant force	F	pounds
	●Radial distance from storm ₊(hurricane) center to region of maximum winds (or to region of maximum waves) (eq. 3-27); (Figure 3-34)	L	nautical miles
	●Distance along bottom contours, as used in refraction problems (R/J method) (Section 2.323)	L	feet
R'	Reduced dynamic component of force per unit wall length from a breaking or broken wave striking structure at oblique angle (eq. 7-103)	F/L	lbs./ft. of wall
R''	Reduced horizontal dynamic component of force per unit wall length from a breaking or broken wave striking nonvertical structure face (eq. 7-104)	F/L	lbs./ft. of wall

Symbol	Definition	Dimension	Example Unit
R_e	Reynolds Number	--	----
R_m	Maximum dynamic component of breaking or broken wave on wall (eq. 7-77)	F/L	lbs./ft. of wall
R_m'	Reduced maximum dynamic component on wall of height lower than wave crest (eq. 7-82)	F/L	lbs./ft. of wall
R_n	Component of R normal to actual wall (Figure 7-81)	F/L	lbs./ft. of wall
R_s	Hydrostatic component of breaking or broken wave on wall (eq. 7-80)	F/L	lbs./ft. of wall
R_t	Total breaking or broken wave force on wall per unit wall length (includes dynamic and hydrostatic components (eq. 7-80)	F/L	lbs./ft. of wall
$R_{\phi\,crit}$	Critical ratio of artificial beach nourishment: ratio of volume required for placement to volume retained on beach after equilibrium (eq. 6-1)	--	----
r	Total rubble layer thickness	L	feet
	●Radial distance from storm (hurricane) center to any specified point in storm system	L	nautical miles
$\sim_r$	Subscript for reflected wave characteristic	--	----
r_A	Armor layer thickness (rubble structure)	L	feet
r_f	Reduction factor for *force* on wall of height lower than clapotis crest (eq. 7-69)	--	----
r_m	Reduction factor for *moment* on wall of height lower than clapotis crest (eq. 7-71)	--	----
	●Reduction factor for maximum *dynamic component* of force when breaking wave height is higher than wall height (eq. 7-82)	--	----
$\sim_{rt}$	Subscript for longshore transport to right as viewed from beach	--	----
r_1	Thickness of first underlayer (rubble structure)	L	feet
S	Channel opening cross-sectional area (eq. 7-111)	L^2	feet^2
	●Surge; height, resulting from storm surge, of free surface above or below the undisturbed water level datum (eq. 3-50); also called wind setup	L	feet

Symbol	Definition	Dimension	Example Unit
Δs	Wave setup between breaker zone and shore (eq. 3-48)	L	feet
	●Wind setup: Difference in water levels at windward and leeward sides of a body of water caused by wind stresses on water surface (eq. 3-82)	L	feet
S_A	Astronomical tide component of total storm surge	L	feet
S_b	Setdown at breaking zone (eq. 3-46)	L	feet
S_D	Dimensionless moment arm of total drag force on pile at given phase angle (eq. 7-29)	—	————
S_{Dm}	Maximum value of S_D	—	————
S_e	Initial setup	L	feet
S_I	Peak surge generated by idealized hurricane moving perpendicular to shoreline at 15 mph. (eq. 3-78)	L	feet
S_i	Dimensionless moment arm of total inertial force on pile at given wave phase angle (eq. 7-28)	—	————
S_{im}	Maximum value of S_i	—	————
S_L	Surge component from water level rise due to local conditions (eq. 3-73)	L	feet
S_o	Observed peak surge (Figure 3-55)	L	feet
S_p	Predicted peak storm surge	L	feet
$S_{\Delta p}$	Component of surge from atmospheric pressure setup	L	feet
S_r	Specific gravity of armor unit (w_r/w_w)	—	————
SSMO	Summary of Synoptic Meteorological Observations	—	————
S_T	Total setup (surge); total water level rise at coast from storm and other causes (eq. 3-73)	L	feet
S_W	Net wave setup at shore (eq. 3-48)	L	feet
S_x	Storm surge component from wind stress perpendicular to coast (eq. 3-62)	L	feet

Symbol	Definition	Dimension	Example Unit
S_y	Storm surge component from wind stress parallel to coast (eq. 3-63)	L	feet
T	Wave period	T	seconds
	● Astronomic tidal period	T	hours
	● Temperature	—	Centigrade or Fahrenheit degrees
T_D	Decayed wave period	T	seconds
T_F	Significant wave period at downwind end of fetch	T	seconds
T_n	Natural, free oscillating period of seiche in closed basin with n nodes (eq. 3-42)	T	hours
T_n'	Free oscillation period in basin open at one end with n' nodes (excluding node at opening) (eq. 3-44)	T	hours
T_o	Deepwater wave period	T	seconds
T_o'	Deepwater wave period corresponding to H_o' used in hurricane wave prediction	T	seconds
T_s	Significant wave period	T	seconds
T_0'	Period of fundamental mode of seiche in rectangular basin open at one end	T	hours
T_1	Fundamental and maximum period of seiches in closed basin	T	hours
t	Time	T	seconds, hours
	● Estimated wind duration over a fetch (Figure 3-15)	T	hours
t_D	Travel time of wave group from end of fetch to a particular location; time or duration of decay	T	hours
t_m	Minimum wind duration over a given fetch for production of a given wave	T	hours
U	Wind speed	L/T	knots, mi./hr.
	● x-component (perpendicular to shore) of volume transport per unit width	L^3/T	mi.^{3}/hr./mi. width
U_g	Geostrophic wind speed (eq. 3-19)	L/T	knots, mi./hr.
U_{gr}	Gradient wind speed (eq. 3-29)	L/T	knots, mi./hr.

Symbol	Definition	Dimension	Example Unit
U_{max}	Maximum gradient wind speed (eq. 3-33)	L/T	knots, mi./hr.
U_R	Maximum sustained gradient wind speed (eq. 3-33) ● Ursell parameter (eq. 2-45)	L/T ——	knots, mi./hr. —————
U_{SM}	Convection term to be added vectorially to wind velocity at each location r to correct for storm motion (eq. 2-13)	L/T	knots, mi./hr.
$\bar{U}(z)$	Mass transport velocity at depth z for a water particle subject to wave motion; mean drift velocity (eq. 2-55)	L/T	ft./sec.
u	Horizontal (x) component of local fluid velocity (water particle velocity); current velocity (eq. 2-13)	L/T	ft./sec.
u_b	Particle velocity under a breaking wave	L/T	ft./sec.
u_{max}	Maximum horizontal water particle velocity	L/T	ft./sec.
$\bar{u}_{max}$	Maximum horizontal water particle velocity averaged over depth	L/T	ft./sec.
V	Velocity ●Maximum velocity of tidal currents in mid-channel (eq. 7-111) ●Volume transport parallel to shore (y-component) (eq. 3-50) ●A volume (eq. 2-65)	L/T L/T L^3/LT L^3/L	knots, mi./hr., ft./sec. ft./sec. mi.^{3}/hr./mi. width ft.^{3}/ft. crest width
V_c	Volume of core (rubble structure)	L^3/L	ft.^{3}/ft.
V_F	Storm center velocity	L/T	mi./hr., knots
V_f	Fall velocity of particles in water column	L/T	ft./sec.
V_1	Volume of first underlayer (rubble structure)	L^3/L	ft.^{3}/ft.
v	Horizontal (y) component of local fluid velocity (water particle velocity); current velocity (Section 3.865a) ●Longshore current velocity	L/T L/T	ft./sec. ft./sec.
v'	Velocity of broken wave water mass at structure located landward of SWL (eq. 7-94)	L/T	ft./sec.
v_b	Longshore current velocity at breaker position	L/T	ft./sec.

Symbol	Definition	Dimension	Example Unit
W	Weight of individual armor unit in primary cover layer; weight of individual units, any layer	F	pounds
	●Fetch width of channel or other restricted body of water (Section 3.432)	L	nautical miles, miles
	●Wind Speed (eq. 3-58)	L/T	knots, mi./hr.
	●Maximum sustained wind speed (Section 3.865b(1))	L/T	knots, mi./hr.
	●Parameter used in pile force and moment calculations (eq. 7-34)	—	————
	●Length of vertical wall affected by unit width of wave crest ($W = 1/\sin \alpha$)	L	feet
W_c	Critical wind speed (eq. 3-58)	L/T	knots, mi./hr.
W_f	Wind speed coefficient (eq. 7-8)	—	————
W_x	x-component of wind speed (eq. 3-50)	L/T	knots, mi./hr.
W_y	y-component of wind speed (eq. 3-51)	L/T	knots, mi./hr.
W_{50}	Weight of 50 percent size of armor riprap gradation (eq. 7-106)	F	pounds
w	Unit weight	F/L^3	lbs./ft.^{3}
	●Vertical (z) component of local fluid velocity or current velocity (eq. 2-14)	L/T	ft./sec.
w_r	Unit weight of armor (rock) unit (saturated surface dry) (eq. 7-110)	F/L^3	lbs./ft.^{3}
w_w	Unit weight of water	F/L^3	lbs./ft.^{3}
X	Distance (Table 3-4)	L	nautical miles
X_i	Distance factor for effective fetch computation (limited bodies of water) (Figure 3-14)	L	miles
$\Delta X_{(z)}$	Net horizontal displacement by water particle z feet below surface during one wave period (eq. 2-55); (Section 2.256)	L	feet
x	Coordinate axis in direction of wave propagation relative to wave crest	—	————
	● Coordinate axis along basin major axis	—	————
	● Coordinate axis perpendicular to and positive toward shore	—	————
	● A distance (eq. 2-10); (eq. 7-50)	L	feet
$\sim_x$	Subscript for x-coordinate	—	————

Symbol	Definition	Dimension	Example Unit
x_n	Location in pile group of nth pile relative to wave crest (eq. 7-49)	L	feet
x_o	Location in pile group of reference pile at wave crest (eq. 7-52)	L	feet
x_p	Plunging breaker travel distance (eq. 7-3)	L	feet
x_r	Location in pile group of reference pile relative to wave crest (eq. 7-52)	L	feet
x_1	Distance from SWL to structure shoreward of SWL (eq. 7-95)	L	feet
x_2	Distance from SWL to limit of wave uprush (eq. 7-95)	L	feet
y	Coordinate axis: *horizontal,* parallel to shore, positive to left when facing shore	––	–––––
	●Coordinate axis: *vertical,* origin at seabed	––	–––––
	●Isbash constant (eq. 7-113)	––	–––––
y_c	Vertical distance from seabed to wave crest (eq. 2-60)	L	feet
y_s	Vertical distance from seabed to water surface (eq. 2-59a)	L	feet
y_t	Vertical distance from seabed to wave trough (eq. 2-60)	L	feet
Z	Time between successive weather charts	T	hours
	●Greenwich mean time	T	hours
z	Coordinate axis: vertical, origin at SWL, positive upwards	––	–––––
$\sim z$	Subscript z refers to z-axis	––	–––––
α (Alpha)	Angle between axis of structure and direction of wave advance (eq. 7-103)	––	degrees
	●Angle between wave crest and bottom contour	––	degrees
	●Angle between wave crest and shore (eq. 2-78a)	––	degrees
	●Angle from wind direction used in determination of effective fetch (Figure 3-14)	––	degrees
	●Upper limit of observed d_b/H_b (Figure 7-2)	––	–––––
	●Empirically determined overtopping coefficient (eq. 7-6)	––	–––––
	●Hurricane movement coefficient (eq. 3-31)	––	–––––
	●Constant for wave spectrum prediction (eq. 3-20)	––	–––––

Symbol	Definition	Dimension	Example Unit
α_b	Angle between breaking wave crest and shore-line	––	degrees
α_m	Coefficient in determination of maximum total moment on pile (eq. 7-36)	––	––––
α_n	Angle, relative to reference pile, that n^{th} pile of pile group makes with direction of wave travel (eq. 7-49)	––	degrees
α_o	Angle between deepwater wave crest and shore-line (eq. 2-78a)	––	degrees
α_x	Local fluid particle acceleration in x-direction (eq. 2-15)	L/T^2	ft./sec.^{2}
α_z	Local fluid particle acceleration in z-direction (eq. 2-16)	L/T^2	ft./sec.^{2}
α_ϕ	Skewness of sediment sample using phi size measures (eq. 4-5)	––	––––
β (Beta)	Angle of beach slope with horizontal ($\tan \beta$ = slope) (eq. 2-87)	––	degrees
	•Constant for wave spectrum prediction (eq. 3-20)	––	––––
	•Lower limit of observed d_b/H_b (Figure 7-2); (eq. 4-15)	––	––––
Γ (Gamma)	Horizontal mixing coefficient in surf zone, perpendicular to shoreline (eq. 4-15)	––	––––
γ	Ratio between left and right longshore transport rates (eq. 4-23)	––	––––
Δ (Delta)	Change; algebraic difference	––	––––
δ	Wall friction angle (eq. 7-116)	––	degrees
ϵ (Epsilon)	Characteristic length describing pile roughness elements (Section 7-311)	L	in., ft.
ζ (Zeta)	Vertical particle displacement caused by wave passage (eq. 2-18)	L	feet
	•Astronomical tide potential in head of water (eq. 3-50)	L	feet
η (Eta)	Displacement of water surface with respect to SWL by passage of wave (eq. 2-10)	L	feet

Symbol	Definition	Dimension	Example Unit
$\eta_{(envelope)}$	Envelope wave form of 2 or more superimposed wave trains (eq. 2-34)	L	feet
η_i	Water surface displacement by incident wave (Section 2.52)	L	feet
η_c	Wave crest elevation above SWL (Section 7.313)	L	feet
η_r	Water surface displacement by reflected wave (Section 2.52)	L	feet
$\eta(t)$	Departure of water surface from its average position as a function of time (eq. 3-11)	L	feet
θ (Theta)	Wave phase angle (Section 2.234)	—	radians
	●Angle of wind measured counterclockwise from x-axis at shore (Section 3.865a); (eq. 3-56)	—	degrees
	●Angle between wind and enclosed basin fetch axis (eq. 3-82)	—	degrees
	●Angle of structure face relative to horizontal (eq. 7-104)	—	degrees
	●Angle of backslope of retaining wall (eq. 7-116)	—	degrees
	●Angle of structure slope in direction of flow (eq. 7-114)	—	degrees
θ_j	Angle between x-axis and direction of propagation of jth wave or wave train (eq. 3-17)	—	degrees
μ (Mu)	Coefficient of friction (soil) (Table 7-14)	—	———
ν (Nu)	Kinematic viscosity (Section 7.311)	L^2/T	ft.^{2}/sec.
ξ (Xi)	Atmospheric pressure deficit in head of water (eq. 3-50)	L	feet
	●Horizontal particle displacement from wave passage (eq. 2-17)	L	feet
π (Pi)	Constant = 3.14159	—	———
ρ (Rho)	Mass density ($\rho = w/g$)	FT^2/L^4	lbs.-sec.^{2}/ft.^{4},(slugs/ft.^{3})
ρ_a	Mass density of air	FT^2/L^4	lbs.-sec.^{2}/ft.^{4},(slugs/ft.^{3})
ρ_{fw}	Mass density of fresh water (1.94 slugs/ft.^{3})	FT^2/L^4	lbs.-sec.^{2}/ft.^{4},(slugs/ft.^{3})
ρ_s	Mass density of sediment	FT^2/L^4	lbs.-sec.^{2}/ft.^{4},(slugs/ft.^{3})

Symbol	Definition	Dimension	Example Unit
ρ_w	Mass density of water (salt water = 2.0 slugs/ft.^{3}; fresh water = 1.94 slugs/ft.^{3})	FT^2/L^4	lb.-sec.^{2}/ft.^{4}, (slugs/ft.^{3})
σ (Sigma)	Standard deviation	––	appropriate units
σ_ϕ	Sediment size standard deviation (phi units)	L	phi
$\sigma_{\phi b}$	Standard deviation of artificial beach nourishment borrow material in phi units (eq. 6-1)	––	phi
$\sigma_{\phi n}$	Standard deviation of native beach material in phi units (eq. 6-1)	––	phi
τ_b (Tau)	Bottom shear stress (eq. 3-79)	F/L^2	lbs./ft.^{2}
τ_p	Dimensionless breaker plunge distance: ratio of breaker travel distance to breaker height	––	–––––
τ_s	Surface shear stress from wind (eq. 3-79)	F/L^2	lbs./ft.^{2}
ϕ (Phi)	Velocity potential	L^2/T	ft.^{2}/sec.
	●Angle between wave direction and plane across which energy is being transmitted (Section 2.238)	––	degrees
	●Angle of incident wave to gap in breakwater	––	degrees
	●Latitude of location	––	degrees
	●Grain size units [$\phi = -\log_2 d(mm)$]	L	phi
	●Internal angle of friction of earth fill or other material	––	degrees
ϕ_j	Phase of jth wave at time t = 0 (eq. 3-11)	––	degrees
ϕ_m	Coefficient for calculation of maximum total force on piles (eq. 7-35)	––	–––––
ϕ_x	Particle size in phi units of the xth percentile in sediment sample	L	phi
χ (Chi)	Wave reflection coefficient (Section 2.51); (eq. 2-85)	––	–––––
χ_1	Wave reflection factor dependent on roughness and permeability of beach, independent of slope (eq. 2-85)	––	–––––
χ_2	Wave reflection factor dependent on beach slope and wave steepness (eq. 2-85)	––	–––––
ψ (Psi)	Angle between storm movement (not wind) direction and coast, measured clockwise from right coast as one looks landward (Figure 3-54)	––	degrees

Symbol	Definition	Dimension	Example Unit
ω (Omega)	Wave angular frequency (eq. 2-3) (eq. 3-20)	T^{-1}	rad./sec.
	● Earth angular frequency (Section 3.4); (eq. 3-19)	T^{-1}	rad./sec., rad./hr.
ω_j	Frequency of jth wave at time t = 0 (eq. 3-11)	T^{-1}	rad./sec.
ω_o	Coefficient for equilibrium wave spectrum calculation (eq. 3-20)	T^{-1}	$second^{-1}$

APPENDIX C

MISCELLANEOUS

TABLES

AND PLATES

APPENDIX C

MISCELLANEOUS TABLES AND PLATES

LIST OF PLATES

LIST OF TABLES

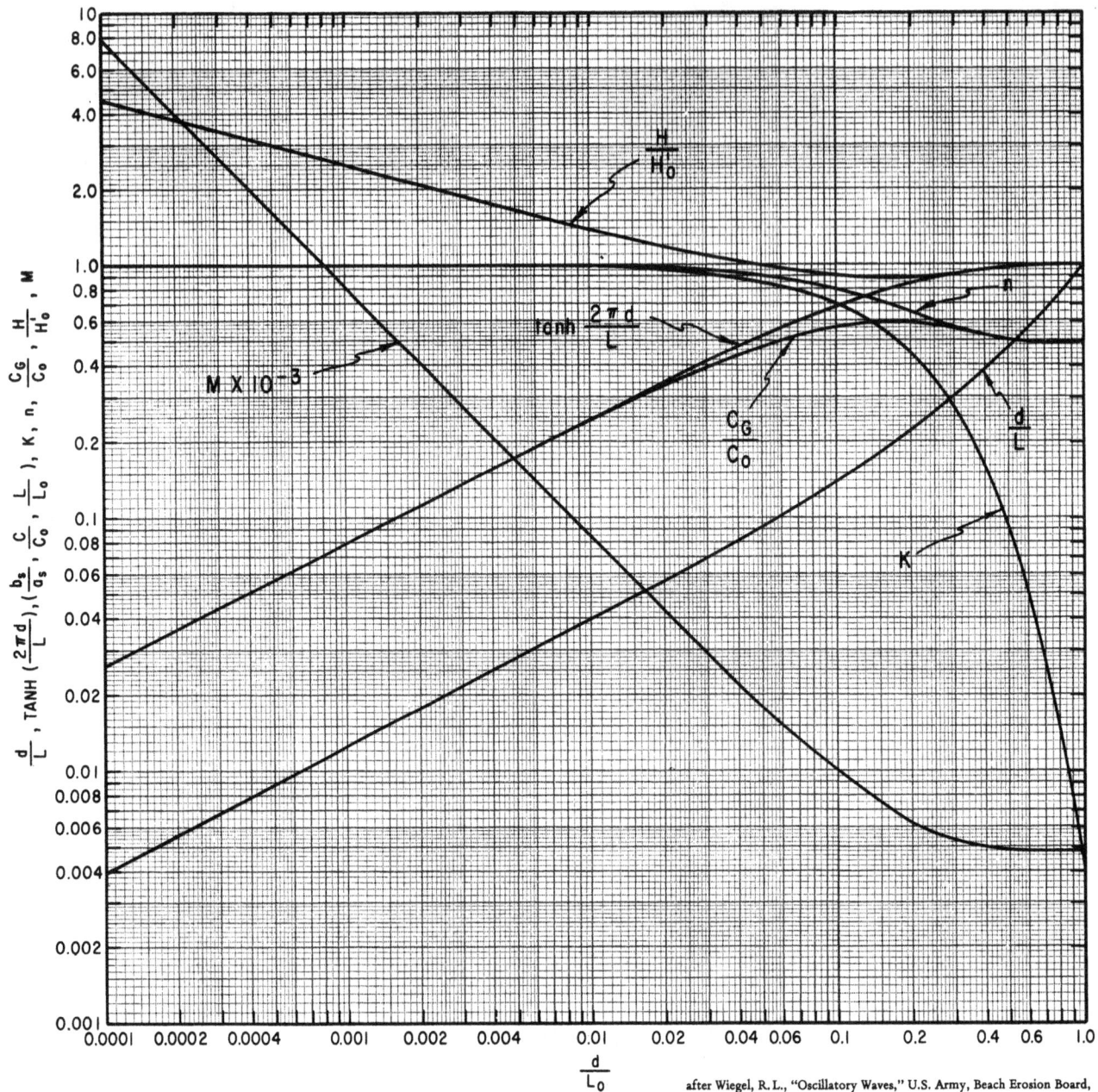

Plate C-1. Illustration of Various Functions of $\frac{d}{L_o}$

after Wiegel, R. L., "Oscillatory Waves," U.S. Army, Beach Erosion Board, Bulletin, Special Issue No. 1, July 1948.

$\dfrac{d}{L_o}$ = ratio of the depth of water at any specific location to the wave length in deep water.

$\dfrac{d}{L}$ = ratio of the depth of water at any specific location to the wave length at that same location.

$\dfrac{H}{H_o'}$ = ratio of the wave height in shallow water to what its wave height would have been in deep water if unaffected by refraction.

$$\frac{H}{H_o'} = \sqrt{\frac{1}{2} \cdot \frac{1}{n} \cdot \frac{1}{C/C_o}} = K_s \text{ (shoaling coefficient)}$$

K = a pressure response factor used in connection with underwater pressure instruments, where

$$K = \frac{H'}{H} = \frac{P}{P_o} = \frac{\cosh\left[2\pi d/L \left(1 + z/d\right)\right]}{\cosh\left(2\pi d/L\right)} \text{ or } \frac{\cosh\left[2\pi (d + z)/L\right]}{\cosh\left(2\pi d/L\right)}$$

where P is the pressure fluctuation at a depth z measured negatively below stillwater, P_o is the surface pressure fluctuation, d is the depth of water from stillwater level to the ocean bottom, L is the wavelength in any particular depth of water, and H is the corresponding variation of head at a depth z. The values of K shown in the tables are for the instrument placed on the bottom using the equation when z = - d.

$$K = \frac{1}{\cosh\left(2\pi d/L\right)} \text{ values tabulated in column 8}$$

n = the fraction of wave energy that travels forward with the wave form: i.e., with the wave velocity C rather than the group velocity C_G.

$$n = \frac{1}{2}\left[1 + \frac{4\pi d/L}{\sinh\left(4\pi d/L\right)}\right] = \frac{C_G}{C}$$

C-3

Guide for Use of Tables C-1 and C-2 -- Continued

n is also the ratio of group velocity C_G to wave velocity C

$\dfrac{C_G}{C_o}$ = ratio of group velocity to deepwater wave velocity where

$$\frac{C_G}{C_o} = \frac{C_G}{C} \times \frac{C}{C_o} = n \, \tanh\left(\frac{2\pi d}{L}\right)$$

M = an energy coefficient defined as

$$\frac{\pi^2}{2 \, \tanh^2\left(2\pi d/_L\right)}$$

Table C–1. Functions of d/L for Even Increments of d/L_o. (from 0.0001 to 1.000)

d/L_o	d/L	$2\pi d/L$	TANH* $2\pi d/L$	SINH $2\pi d/L$	COSH $2\pi d/L$	H/H'_o	K	$4\pi d/L$	SINH $4\pi d/L$	COSH $4\pi d/L$	n	C_G/C_o	M
0	0	0	0	0	1	∝	1	0	0	1	1	0	∝
.0001000	.003990	.02507	.02506	.02507	1.0003	4.467	.9997	.05014	.05016	1.001	.9998	.02506	7,855
.0002000	.005643	.03546	.03544	.03547	1.0006	3.757	.9994	.07091	.07097	1.003	.9996	.03543	3,928
.0003000	.006912	.04343	.04340	.04344	1.0009	3.395	.9991	.08686	.08697	1.004	.9994	.04336	2,620
.0004000	.007982	.05015	.05011	.05018	1.0013	3.160	.9987	.1003	.1005	1.005	.9992	.05007	1,965
.0005000	.008925	.05608	.05602	.05611	1.0016	2.989	.9984	.1122	.1124	1.006	.9990	.05596	1,572
.0006000	.009778	.06144	.06136	.06148	1.0019	2.856	.9981	.1229	.1232	1.008	.9988	.06128	1,311
.0007000	.01056	.06637	.06627	.06642	1.0022	2.749	.9978	.1327	.1331	1.009	.9985	.06617	1,124
.0008000	.01129	.07096	.07084	.07102	1.0025	2.659	.9975	.1419	.1424	1.010	.9983	.07072	983.5
.0009000	.01198	.07527	.07513	.07534	1.0028	2.582	.9972	.1505	.1511	1.011	.9981	.07499	874.3
.001000	.01263	.07935	.07918	.07943	1.0032	2.515	.9969	.1587	.1594	1.013	.9979	.07902	787.0
.001100	.01325	.08323	.08304	.08333	1.0035	2.456	.9966	.1665	.1672	1.014	.9977	.08285	715.6
.001200	.01384	.08694	.08672	.08705	1.0038	2.404	.9962	.1739	.1748	1.015	.9975	.08651	656.1
.001300	.01440	.09050	.09026	.09063	1.0041	2.357	.9959	.1810	.1820	1.016	.9973	.09001	605.8
.001400	.01495	.09393	.09365	.09407	1.0044	2.314	.9956	.1879	.1890	1.018	.9971	.09338	562.6
.001500	.01548	.09723	.09693	.09739	1.0047	2.275	.9953	.1945	.1957	1.019	.9969	.09663	525
.001600	.01598	.1004	.1001	.1006	1.0051	2.239	.9949	.2009	.2022	1.020	.9967	.09977	493
.001700	.01648	.1035	.1032	.1037	1.0054	2.205	.9946	.2071	.2086	1.022	.9965	.1028	463
.001800	.01696	.1066	.1062	.1068	1.0057	2.174	.9943	.2131	.2147	1.023	.9962	.1058	438
.001900	.01743	.1095	.1091	.1097	1.0060	2.145	.9940	.2190	.2207	1.024	.9960	.1087	415
.002000	.01788	.1123	.1119	.1125	1.0063	2.119	.9937	.2247	.2266	1.025	.9958	.1114	394
.002100	.01832	.1151	.1146	.1154	1.0066	2.094	.9934	.2303	.2323	1.027	.9956	.1141	376
.002200	.01876	.1178	.1173	.1181	1.0069	2.070	.9931	.2357	.2379	1.028	.9954	.1161	359
.002300	.01918	.1205	.1199	.1208	1.0073	2.047	.9928	.2410	.2433	1.029	.9952	.1193	343
.002400	.01959	.1231	.1225	.1234	1.0076	2.025	.9925	.2462	.2487	1.031	.9950	.1219	329
.002500	.02000	.1257	.1250	.1260	1.0079	2.005	.9922	.2513	.2540	1.032	.9948	.1243	316
.002600	.02040	.1282	.1275	.1285	1.0082	1.986	.9919	.2563	.2592	1.033	.9946	.1268	304
.002700	.02079	.1306	.1299	.1310	1.0085	1.967	.9916	.2612	.2642	1.034	.9944	.1292	292
.002800	.02117	.1330	.1323	.1334	1.0089	1.950	.9912	.2661	.2692	1.036	.9942	.1315	282
.002900	.02155	.1354	.1346	.1358	1.0092	1.933	.9909	.2708	.2741	1.037	.9939	.1338	272
.003000	.02192	.1377	.1369	.1382	1.0095	1.917	.9906	.2755	.2790	1.038	.9937	.1360	263
.003100	.02228	.1400	.1391	.1405	1.0098	1.902	.9903	.2800	.2837	1.040	.9935	.1382	255
.003200	.02264	.1423	.1413	.1427	1.0101	1.887	.9900	.2845	.2884	1.041	.9933	.1404	247
.003300	.02300	.1445	.1435	.1449	1.0104	1.873	.9897	.2890	.2930	1.042	.9931	.1425	240
.003400	.02335	.1467	.1456	.1472	1.0108	1.860	.9893	.2934	.2976	1.043	.9929	.1446	233
.003500	.02369	.1488	.1477	.1494	1.0111	1.847	.9890	.2977	.3021	1.045	.9927	.1466	226
.003600	.02403	.1510	.1498	.1515	1.0114	1.834	.9887	.3020	.3065	1.046	.9925	.1487	220
.003700	.02436	.1531	.1519	.1537	1.0117	1.822	.9884	.3061	.3109	1.047	.9923	.1507	214
.003800	.02469	.1551	.1539	.1558	1.0121	1.810	.9881	.3103	.3153	1.049	.9921	.1527	208
.003900	.02502	.1572	.1559	.1579	1.0124	1.799	.9878	.3144	.3196	1.050	.9919	.1546	203
.004000	.02534	.1592	.1579	.1599	1.0127	1.788	.9875	.3184	.3238	1.051	.9917	.1565	198
.004100	.02566	.1612	.1598	.1619	1.0130	1.777	.9872	.3224	.3280	1.052	.9915	.1584	193
.004200	.02597	.1632	.1617	.1639	1.0133	1.767	.9869	.3263	.3322	1.054	.9912	.1602	189
.004300	.02628	.1651	.1636	.1659	1.0137	1.756	.9865	.3302	.3362	1.055	.9910	.1621	184
.004400	.02659	.1671	.1655	.1678	1.0140	1.746	.9862	.3341	.3403	1.056	.9908	.1640	180
.004500	.02689	.1690	.1674	.1698	1.0143	1.737	.9859	.3380	.3444	1.058	.9906	.1658	176
.004600	.02719	.1708	.1692	.1717	1.0146	1.727	.9856	.3417	.3483	1.059	.9904	.1676	172
.004700	.02749	.1727	.1710	.1736	1.0149	1.718	.9853	.3454	.3523	1.060	.9902	.1693	169
.004800	.02778	.1745	.1728	.1754	1.0153	1.709	.9849	.3491	.3562	1.062	.9900	.1711	165
.004900	.02807	.1764	.1746	.1773	1.0156	1.701	.9846	.3527	.3601	1.063	.9898	.1728	162
.005000	.02836	.1782	.1764	.1791	1.0159	1.692	.9843	.3564	.3640	1.064	.9896	.1746	159
.005100	.02864	.1800	.1781	.1809	1.0162	1.684	.9840	.3599	.3678	1.066	.9894	.1762	156
.005200	.02893	.1818	.1798	.1827	1.0166	1.676	.9837	.3635	.3715	1.067	.9892	.1779	153
.005300	.02921	.1835	.1815	.1845	1.0169	1.669	.9834	.3670	.3753	1.068	.9889	.1795	150
.005400	.02948	.1852	.1832	.1863	1.0172	1.662	.9831	.3705	.3790	1.069	.9887	.1811	147
.005500	.02976	.1870	.1848	.1880	1.0175	1.654	.9828	.3739	.3827	1.071	.9885	.1827	145
.005600	.03003	.1887	.1865	.1898	1.0178	1.647	.9825	.3774	.3864	1.072	.9883	.1843	142
.005700	.03030	.1904	.1881	.1915	1.0182	1.640	.9822	.3808	.3900	1.073	.9881	.1859	140
.005800	.03057	.1921	.1897	.1932	1.0185	1.633	.9818	.3841	.3937	1.075	.9879	.1874	137
.005900	.03083	.1937	.1913	.1949	1.0188	1.626	.9815	.3875	.3972	1.076	.9877	.1890	135

*Also: b_s/a_s, C/C_o, L/L_o

C-5

Table C—1 — Continued

d/L$_o$	d/L	2πd/L	TANH 2πd/L	SINH 2πd/L	COSH 2πd/L	H/H$'_o$	K	4πd/L	SINH 4πd/L	COSH 4πd/L	n	C$_G$/C$_o$	M
.006000	.03110	.1954	.1929	.1967	1.0192	1.620	.9812	.3908	.4008	1.077	.9875	.1905	133
.006100	.03136	.1970	.1945	.1983	1.0195	1.614	.9809	.3941	.4044	1.079	.9873	.1920	130
.006200	.03162	.1987	.1961	.2000	1.0198	1.607	.9806	.3973	.4079	1.080	.9871	.1935	128
.006300	.03188	.2003	.1976	.2016	1.0201	1.601	.9803	.4006	.4114	1.081	.9869	.1950	126
.006400	.03213	.2019	.1992	.2033	1.0205	1.595	.9799	.4038	.4148	1.083	.9867	.1965	124
.006500	.03238	.2035	.2007	.2049	1.0208	1.589	.9796	.4070	.4183	1.084	.9865	.1980	123
.006600	.03264	.2051	.2022	.2065	1.0211	1.583	.9793	.4101	.4217	1.085	.9863	.1994	121
.006700	.03289	.2066	.2037	.2081	1.0214	1.578	.9790	.4133	.4251	1.087	.9860	.2009	119
.006800	.03313	.2082	.2052	.2097	1.0217	1.572	.9787	.4164	.4285	1.088	.9858	.2023	117
.006900	.03338	.2097	.2067	.2113	1.0221	1.567	.9784	.4195	.4319	1.089	.9856	.2037	116
.007000	.03362	.2113	.2082	.2128	1.0224	1.561	.9781	.4225	.4352	1.091	.9854	.2051	114
.007100	.03387	.2128	.2096	.2144	1.0227	1.556	.9778	.4256	.4386	1.092	.9852	.2065	112
.007200	.03411	.2143	.2111	.2160	1.0231	1.551	.9774	.4286	.4419	1.093	.9850	.2079	111
.007300	.03435	.2158	.2125	.2175	1.0234	1.546	.9771	.4316	.4452	1.095	.9848	.2093	109
.007400	.03459	.2173	.2139	.2190	1.0237	1.541	.9768	.4346	.4484	1.096	.9846	.2106	108
.007500	.03482	.2188	.2154	.2205	1.0240	1.536	.9765	.4376	.4517	1.097	.9844	.2120	106
.007600	.03506	.2203	.2168	.2221	1.0244	1.531	.9762	.4406	.4549	1.099	.9842	.2134	105
.007700	.03529	.2218	.2182	.2236	1.0247	1.526	.9759	.4435	.4582	1.100	.9840	.2147	104
.007800	.03552	.2232	.2196	.2251	1.0250	1.521	.9756	.4464	.4614	1.101	.9838	.2160	102
.007900	.03576	.2247	.2209	.2265	1.0253	1.517	.9753	.4493	.4646	1.103	.9836	.2173	101
.008000	.03598	.2261	.2223	.2280	1.0257	1.512	.9750	.4522	.4678	1.104	.9834	.2186	100
.008100	.03621	.2275	.2237	.2295	1.0260	1.508	.9747	.4551	.4709	1.105	.9832	.2199	98.6
.008200	.03644	.2290	.2250	.2310	1.0263	1.503	.9744	.4579	.4741	1.107	.9830	.2212	97.5
.008300	.03666	.2304	.2264	.2324	1.0266	1.499	.9741	.4607	.4772	1.108	.9827	.2225	96.3
.008400	.03689	.2318	.2277	.2338	1.0270	1.495	.9737	.4636	.4803	1.109	.9825	.2237	95.2
.008500	.03711	.2332	.2290	.2353	1.0273	1.491	.9734	.4664	.4834	1.111	.9823	.2250	94.1
.008600	.03733	.2346	.2303	.2367	1.0276	1.487	.9731	.4691	.4865	1.112	.9821	.2262	93.0
.008700	.03755	.2360	.2317	.2381	1.0280	1.482	.9728	.4719	.4896	1.113	.9819	.2275	91.9
.008800	.03777	.2373	.2330	.2396	1.0283	1.478	.9725	.4747	.4927	1.115	.9817	.2287	90.9
.008900	.03799	.2387	.2343	.2410	1.0286	1.474	.9722	.4774	.4957	1.116	.9815	.2300	89.9
.009000	.03821	.2401	.2356	.2424	1.0290	1.471	.9718	.4801	.4988	1.118	.9813	.2312	88.9
.009100	.03842	.2414	.2368	.2438	1.0293	1.467	.9715	.4828	.5018	1.119	.9811	.2324	88.0
.009200	.03864	.2428	.2381	.2452	1.0296	1.463	.9712	.4855	.5049	1.120	.9809	.2336	87.1
.009300	.03885	.2441	.2394	.2465	1.0299	1.459	.9709	.4882	.5079	1.122	.9807	.2348	86.1
.009400	.03906	.2455	.2407	.2479	1.0303	1.456	.9706	.4909	.5109	1.123	.9805	.2360	85.2
.009500	.03928	.2468	.2419	.2493	1.0306	1.452	.9703	.4936	.5138	1.124	.9803	.2371	84.3
.009600	.03949	.2481	.2431	.2507	1.0309	1.448	.9700	.4962	.5168	1.126	.9801	.2383	83.5
.009700	.03970	.2494	.2443	.2520	1.0313	1.445	.9697	.4988	.5198	1.127	.9799	.2394	82.7
.009800	.03990	.2507	.2456	.2534	1.0316	1.442	.9694	.5014	.5227	1.128	.9797	.2406	81.8
.009900	.04011	.2520	.2468	.2547	1.0319	1.438	.9691	.5040	.5257	1.130	.9794	.2417	81.0
.01000	.04032	.2533	.2480	.2560	1.0322	1.435	.9688	.5066	.5286	1.131	.9792	.2429	80.2
.01100	.04233	.2660	.2598	.2691	1.0356	1.403	.9656	.5319	.5574	1.145	.9772	.2539	73.1
.01200	.04426	.2781	.2711	.2817	1.0389	1.375	.9625	.5562	.5853	1.159	.9751	.2643	67.1
.01300	.04612	.2898	.2820	.2938	1.0423	1.350	.9594	.5795	.6125	1.173	.9731	.2743	62.1
.01400	.04791	.3010	.2924	.3056	1.0456	1.327	.9564	.6020	.6391	1.187	.9710	.2838	57.8
.01500	.04964	.3119	.3022	.3170	1.0490	1.307	.9533	.6238	.6651	1.201	.9690	.2928	54.0
.01600	.05132	.3225	.3117	.3281	1.0524	1.288	.9502	.6450	.6906	1.215	.9670	.3014	50.8
.01700	.05296	.3328	.3209	.3389	1.0559	1.271	.9471	.6655	.7158	1.230	.9649	.3096	47.9
.01800	.05455	.3428	.3298	.3495	1.0593	1.255	.9440	.6856	.7405	1.244	.9629	.3176	45.3
.01900	.05611	.3525	.3386	.3599	1.0628	1.240	.9409	.7051	.7650	1.259	.9609	.3253	43.0
.02000	.05763	.3621	.3470	.3701	1.0663	1.226	.9378	.7242	.7891	1.274	.9588	.3327	41.0
.02100	.05912	.3714	.3552	.3800	1.0698	1.213	.9348	.7429	.8131	1.289	.9568	.3399	39.1
.02200	.06057	.3806	.3632	.3898	1.0733	1.201	.9317	.7612	.8368	1.304	.9548	.3468	37.4
.02300	.06200	.3896	.3710	.3995	1.0768	1.189	.9287	.7791	.8603	1.319	.9528	.3535	35.9
.02400	.06340	.3984	.3786	.4090	1.0804	1.178	.9256	.7967	.8837	1.335	.9508	.3600	34.4
.02500	.06478	.4070	.3860	.4184	1.0840	1.168	.9225	.8140	.9069	1.350	.9488	.3662	33.1
.02600	.06613	.4155	.3932	.4276	1.0876	1.159	.9195	.8310	.9310	1.366	.9468	.3722	31.9
.02700	.06747	.4239	.4002	.4367	1.0912	1.150	.9164	.8478	.9530	1.381	.9448	.3781	30.8
.02800	.06878	.4322	.4071	.4457	1.0949	1.141	.9133	.8643	.9760	1.397	.9428	.3838	29.8
.02900	.07007	.4403	.4138	.4546	1.0985	1.133	.9103	.8805	.9988	1.413	.9408	.3893	28.8

Table C—1 — Continued

d/L$_o$	d/L	2πd/L	TANH 2πd/L	SINH 2πd/L	COSH 2πd/L	H/H'$_o$	K	4πd/L	SINH 4πd/L	COSH 4πd/L	n	C$_G$/C$_O$	M
.03000	.07135	.4483	.4205	.4634	1.1021	1.125	.9073	.8966	1.022	1.430	.9388	.3947	27.9
.03100	.07260	.4562	.4269	.4721	1.1059	1.118	.9042	.9124	1.044	1.446	.9369	.4000	27.1
.03200	.07385	.4640	.4333	.4808	1.1096	1.111	.9012	.9280	1.067	1.462	.9349	.4051	26.3
.03300	.07507	.4717	.4395	.4894	1.1133	1.104	.8982	.9434	1.090	1.479	.9329	.4100	25.6
.03400	.07630	.4794	.4457	.4980	1.1171	1.098	.8952	.9588	1.113	1.496	.9309	.4149	24.8
.03500	.07748	.4868	.4517	.5064	1.1209	1.092	.8921	.9737	1.135	1.513	.9289	.4196	24.19
.03600	.07867	.4943	.4577	.5147	1.1247	1.086	.8891	.9886	1.158	1.530	.9270	.4242	23.56
.03700	.07984	.5017	.4635	.5230	1.1285	1.080	.8861	1.0033	1.180	1.547	.9250	.4287	22.97
.03800	.08100	.5090	.4691	.5312	1.1324	1.075	.8831	1.018	1.203	1.564	.9230	.4330	22.42
.03900	.08215	.5162	.4747	.5394	1.1362	1.069	.8801	1.032	1.226	1.582	.9211	.4372	21.90
.04000	.08329	.5233	.4802	.5475	1.1401	1.064	.8771	1.047	1.248	1.600	.9192	.4414	21.40
.04100	.08442	.5304	.4857	.5556	1.1440	1.059	.8741	1.061	1.271	1.617	.9172	.4455	20.92
.04200	.08553	.5374	.4911	.5637	1.1479	1.055	.8711	1.075	1.294	1.636	.9153	.4495	20.46
.04300	.08664	.5444	.4964	.5717	1.1518	1.050	.8688	1.089	1.317	1.654	.9133	.4534	20.03
.04400	.08774	.5513	.5015	.5796	1.1558	1.046	.8652	1.103	1.340	1.672	.9114	.4571	19.62
.04500	.08883	.5581	.5066	.5876	1.1599	1.042	.8621	1.116	1.363	1.691	.9095	.4607	19.23
.04600	.08991	.5649	.5116	.5954	1.1639	1.038	.8592	1.130	1.386	1.709	.9076	.4643	18.85
.04700	.09098	.5717	.5166	.6033	1.1679	1.034	.8562	1.143	1.409	1.728	.9057	.4679	18.49
.04800	.09205	.5784	.5215	.6111	1.1720	1.030	.8532	1.157	1.433	1.747	.9037	.4713	18.15
.04900	.09311	.5850	.5263	.6189	1.1760	1.026	.8503	1.170	1.456	1.766	.9018	.4746	17.82
05000	.09416	.5916	.5310	.6267	1.1802	1.023	.8473	1.183	1.479	1.786	.8999	.4779	17.50
05100	.09520	.5981	.5357	.6344	1.1843	1.019	.8444	1.196	1.503	1.805	.8980	.4811	17.19
05200	.09623	.6046	.5403	.6421	1.1884	1.016	.8415	1.209	1.526	1.825	.8961	.4842	16.90
05300	.09726	.6111	.5449	.6499	1.1926	1.013	.8385	1.222	1.550	1.845	.8943	.4873	16.62
05400	.09829	.6176	.5494	.6575	1.1968	1.010	.8356	1.235	1.574	1.865	.8924	.4903	16.35
05500	.09930	.6239	.5538	.6652	1.2011	1.007	.8326	1.248	1.598	1.885	.8905	.4932	16.09
05600	.1003	.6303	.5582	.6729	1.2053	1.004	.8297	1.261	1.622	1.906	.8886	.4960	15.84
05700	.1013	.6366	.5626	.6805	1.2096	1.001	.8267	1.273	1.646	1.926	.8867	.4988	15.60
05800	.1023	.6428	.5668	.6880	1.2138	.9985	.8239	1.286	1.670	1.947	.8849	.5015	15.36
05900	.1033	.6491	.5711	.6956	1.2181	.9958	.8209	1.298	1.695	1.968	.8830	.5042	15.13
.06000	.1043	.6553	.5753	.7033	1.2225	.9932	.8180	1.311	1.719	1.989	.8811	.5068	14.91
.06100	.1053	.6616	.5794	.7110	1.2270	.9907	.8150	1.3231	1.744	2.011	.8792	.5094	14.70
.06200	.1063	.6678	.5834	.7187	1.2315	.9883	.8121	1.336	1.770	2.033	.8773	.5119	14.50
.06300	.1073	.6739	.5874	.7256	1.2355	.9860	.8093	1.348	1.795	2.055	.8755	.5143	14.30
.06400	.1082	.6799	.5914	.7335	1.2402	.9837	.8063	1.360	1.819	2.076	.8737	.5167	14.11
.06500	.1092	.6860	.5954	.7411	1.2447	.9815	.8035	1.372	1.845	2.098	.8719	.5191	13.92
.06600	.1101	.6920	.5993	.7486	1.2492	.9793	.8005	1.384	1.870	2.121	.8700	.5214	13.74
.06700	.1111	.6981	.6031	.7561	1.2537	.9772	.7977	1.396	1.896	2.144	.8682	.5236	13.57
.06800	.1120	.7037	.6069	.7633	1.2580	.9752	.7948	1.408	1.921	2.166	.8664	.5258	13.40
.06900	.1130	.7099	.6106	.7711	1.2628	.9732	.7919	1.420	1.948	2.189	.8646	.5279	13.24
.07000	.1139	.7157	.6144	.7783	1.2672	.9713	.7890	1.432	1.974	2.213	.8627	.5300	13.08
.07100	.1149	.7219	.6181	.7863	1.2721	.9694	.7861	1.444	2.000	2.236	.8609	.5321	12.92
.07200	.1158	.7277	.6217	.7937	1.2767	.9676	.7833	1.455	2.026	2.260	.8591	.5341	12.77
.07300	.1168	.7336	.6252	.8011	1.2813	.9658	.7804	1.467	2.053	2.284	.8572	.5360	12.62
.07400	.1177	.7395	.6289	.8088	1.2861	.9641	.7775	1.479	2.080	2.308	.8554	.5380	12.48
.07500	.1186	.7453	.6324	.8162	1.2908	.9624	.7747	1.490	2.107	2.332	.8537	.5399	12.34
07600	.1195	.7511	.6359	.8237	1.2956	.9607	.7719	1.502	2.135	2.357	.8519	.5417	12.21
07700	.1205	.7569	.6392	.8312	1.3004	.9591	.7690	1.514	2.162	2.382	.8501	.5435	12.08
07800	.1214	.7625	.6427	.8386	1.3051	.9576	.7662	1.525	2.189	2.407	.8483	.5452	11.95
07900	.1223	.7683	.6460	.8462	1.3100	.9562	.7634	1.537	2.217	2.432	.8465	.5469	11.83
08000	.1232	.7741	.6493	.8538	1.3149	.9548	.7605	1.548	2.245	2.458	.8448	.5485	11.71
08100	.1241	.7799	.6526	.8614	1.3198	.9534	.7577	1.560	2.274	2.484	.8430	.5501	11.59
08200	.1251	.7854	.6558	.8687	1.3246	.9520	.7549	1.571	2.303	2.511	.8413	.5517	11.47
08300	.1259	.7911	.6590	.8762	1.3295	.9506	.7522	1.583	2.331	2.537	.8395	.5533	11.36
08400	.1268	.7967	.6622	.8837	1.3345	.9493	.7494	1.594	2.360	2.563	.8378	.5548	11.25
08500	.1277	.8026	.6655	.8915	1.3397	.9481	.7464	1.605	2.389	2.590	.8360	.5563	11.14
08600	.1286	.8080	.6685	.8989	1.3446	.9469	.7437	1.616	2.418	2.617	.8342	.5577	11.04
08700	.1295	.8137	.6716	.9064	1.3497	.9457	.7409	1.628	2.448	2.644	.8325	.5591	10.94
08800	.1304	.8193	.6747	.9141	1.3548	.9445	.7381	1.639	2.478	2.672	.8308	.5605	10.84
08900	.1313	.8250	.6778	.9218	1.3600	.9433	.7353	1.650	2.508	2.700	.8290	.5619	10.74

Table C–1 – Continued

d/L_0	d/L	$2\pi d/L$	TANH $2\pi d/L$	SINH $2\pi d/L$	COSH $2\pi d/L$	H/H'_0	K	$4\pi d/L$	SINH $4\pi d/L$	COSH $4\pi d/L$	n	c_G/c_0	M
.09000	.1322	.8306	.6808	.9295	1.3653	.9422	.7324	1.661	2.538	2.728	.8273	.5632	10.65
.09100	.1331	.8363	.6838	.9372	1.3706	.9411	.7296	1.672	2.568	2.756	.8255	.5645	10.55
.09200	.1340	.8420	.6868	.9450	1.3759	.9401	.7268	1.684	2.599	2.785	.8238	.5658	10.46
.09300	.1349	.8474	.6897	.9525	1.3810	.9391	.7241	1.695	2.630	2.814	.8221	.5670	10.37
.09400	.1357	.8528	.6925	.9600	1.3862	.9381	.7214	1.706	2.662	2.843	.8204	.5682	10.29
.09500	.1366	.8583	.6953	.9677	1.3917	.9371	.7186	1.717	2.693	2.873	.8187	.5693	10.21
.09600	.1375	.8639	.6982	.9755	1.3970	.9362	.7158	1.728	2.726	2.903	.8170	.5704	10.12
.09700	.1384	.8694	.7011	.9832	1.4023	.9353	.7131	1.739	2.757	2.933	.8153	.5716	10.04
.09800	.1392	.8749	.7039	.9908	1.4077	.9344	.7104	1.750	2.790	2.963	.8136	.5727	9.962
.09900	.1401	.8803	.7066	.9985	1.4131	.9335	.7076	1.761	2.822	2.994	.8120	.5737	9.884
.1000	.1410	.8858	.7093	1.006	1.4187	.9327	.7049	1.772	2.855	3.025	.8103	.5747	9.808
.1010	.1419	.8913	.7120	1.014	1.4242	.9319	.7022	1.783	2.888	3.057	.8086	.5757	9.734
.1020	.1427	.8967	.7147	1.022	1.4297	.9311	.6994	1.793	2.922	3.088	.8069	.5766	9.661
.1030	.1436	.9023	.7173	1.030	1.4354	.9304	.6967	1.805	2.956	3.121	.8052	.5776	9.590
.1040	.1445	.9076	.7200	1.037	1.4410	.9297	.6940	1.815	2.990	3.153	.8036	.5785	9.519
.1050	.1453	.9130	.7226	1.045	1.4465	.9290	.6913	1.826	3.024	3.185	.8019	.5794	9.451
.1060	.1462	.9184	.7252	1.053	1.4523	.9282	.6886	1.837	3.059	3.218	.8003	.5803	9.384
.1070	.1470	.9239	.7277	1.061	1.4580	.9276	.6859	1.848	3.094	3.251	.7986	.5812	9.318
.1080	.1479	.9293	.7303	1.069	1.4638	.9269	.6833	1.858	3.128	3.284	.7970	.5820	9.254
.1090	.1488	.9343	.7327	1.076	1.4692	.9263	.6806	1.869	3.164	3.319	.7954	.5828	9.191
.1100	.1496	.9400	.7352	1.085	1.4752	.9257	.6779	1.880	3.201	3.353	.7937	.5836	9.129
.1110	.1505	.9456	.7377	1.093	1.4814	.9251	.6752	1.891	3.237	3.388	.7920	.5843	9.068
.1120	.1513	.9508	.7402	1.101	1.4871	.9245	.6725	1.902	3.274	3.423	.7904	.5850	9.009
.1130	.1522	.9563	.7426	1.109	1.4932	.9239	.6697	1.913	3.312	3.459	.7888	.5857	8.950
.1140	.1530	.9616	.7450	1.117	1.4990	.9234	.6671	1.923	3.348	3.494	.7872	.5864	8.891
.1150	.1539	.9670	.7474	1.125	1.5051	.9228	.6645	1.934	3.385	3.530	.7856	.5871	8.835
.1160	.1547	.9720	.7497	1.133	1.5108	.9223	.6619	1.944	3.423	3.566	.7840	.5878	8.780
.1170	.1556	.9775	.7520	1.141	1.5171	.9218	.6592	1.955	3.462	3.603	.7824	.5884	8.726
.1180	.1564	.9827	.7543	1.149	1.5230	.9214	.6566	1.966	3.501	3.641	.7808	.5890	8.673
.1190	.1573	.9882	.7566	1.157	1.5293	.9209	.6539	1.977	3.540	3.678	.7792	.5896	8.621
.1200	.1581	.9936	.7589	1.165	1.5356	.9204	.6512	1.987	3.579	3.716	.7776	.5902	8.569
.1210	.1590	.9989	.7612	1.174	1.5418	.9200	.6486	1.998	3.620	3.755	.7760	.5907	8.518
.1220	.1598	1.004	.7634	1.182	1.5479	.9196	.6460	2.008	3.659	3.793	.7745	.5913	8.468
.1230	.1607	1.010	.7656	1.190	1.5546	.9192	.6433	2.019	3.699	3.832	.7729	.5918	8.419
.1240	.1615	1.015	.7678	1.198	1.5605	.9189	.6407	2.030	3.740	3.871	.7713	.5922	8.371
.1250	.1624	1.020	.7700	1.207	1.5674	.9186	.6381	2.041	3.782	3.912	.7698	.5926	8.324
.1260	.1632	1.025	.7721	1.215	1.5734	.9182	.6356	2.051	3.824	3.952	.7682	.5931	8.278
.1270	.1640	1.030	.7742	1.223	1.5795	.9178	.6331	2.061	3.865	3.992	.7667	.5936	8.233
.1280	.1649	1.036	.7763	1.231	1.5862	.9175	.6305	2.072	3.907	4.033	.7652	.5940	8.189
.1290	.1657	1.041	.7783	1.240	1.5927	.9172	.6279	2.082	3.950	4.074	.7637	.5944	8.146
.1300	.1665	1.046	.7804	1.248	1.5990	.9169	.6254	2.093	3.992	4.115	.7621	.5948	8.103
.1310	.1674	1.052	.7824	1.257	1.6060	.9166	.6228	2.104	4.036	4.158	.7606	.5951	8.061
.1320	.1682	1.057	.7844	1.265	1.6124	.9164	.6202	2.114	4.080	4.201	.7591	.5954	8.020
.1330	.1691	1.062	.7865	1.273	1.6191	.9161	.6176	2.125	4.125	4.245	.7575	.5958	7.978
.1340	.1699	1.068	.7885	1.282	1.6260	.9158	.6150	2.135	4.169	4.288	.7560	.5961	7.937
.1350	.1708	1.073	.7905	1.291	1.633	.9156	.6123	2.146	4.217	4.334	.7545	.5964	7.897
.1360	.1716	1.078	.7925	1.300	1.640	.9154	.6098	2.156	4.262	4.378	.7530	.5967	7.857
.1370	.1724	1.084	.7945	1.308	1.647	.9152	.6073	2.167	4.309	4.423	.7515	.5969	7.819
.1380	.1733	1.089	.7964	1.317	1.654	.9150	.6047	2.177	4.355	4.468	.7500	.5972	7.781
.1390	.1741	1.094	.7983	1.326	1.660	.9148	.6022	2.188	4.402	4.514	.7485	.5975	7.744
.1400	.1749	1.099	.8002	1.334	1.667	.9146	.5998	2.198	4.450	4.561	.7471	.5978	7.707
.1410	.1758	1.105	.8021	1.343	1.675	.9144	.5972	2.209	4.498	4.607	.7456	.5980	7.671
.1420	.1766	1.110	.8039	1.352	1.681	.9142	.5947	2.219	4.546	4.654	.7441	.5982	7.636
.1430	.1774	1.115	.8057	1.360	1.688	.9141	.5923	2.230	4.595	4.663	.7426	.5984	7.602
.1440	.1783	1.120	.8076	1.369	1.696	.9140	.5898	2.240	4.644	4.751	.7412	.5986	7.567
.1450	.1791	1.125	.8094	1.378	1.703	.9139	.5873	2.251	4.695	4.800	.7397	.5987	7.533
.1460	.1800	1.131	.8112	1.388	1.710	.9137	.5847	2.261	4.746	4.850	.7382	.5989	7.499
.1470	.1808	1.136	.8131	1.397	1.718	.9136	.5822	2.272	4.798	4.901	.7368	.5990	7.465
.1480	.1816	1.141	.8149	1.405	1.725	.9135	.5798	2.282	4.847	4.951	.7354	.5992	7.432
.1490	.1825	1.146	.8166	1.415	1.732	.9134	.5773	2.293	4.901	5.001	.7339	.5993	7.400

Table C–1 – Continued

d/L_o	d/L	2π d/L	TANH 2π d/L	SINH 2π d/L	COSH 2π d/L	H/H'_o	K	4π d/L	SINH 4π d/L	COSH 4π d/L	n	C_G/C_o	M
.1500	.1833	1.152	.8183	1.424	1.740	.9133	.5748	2.303	4.954	5.054	.7325	.5994	7.369
.1510	.1841	1.157	.8200	1.433	1.747	.9133	.5723	2.314	5.007	5.106	.7311	.5994	7.339
.1520	.1850	1.162	.8217	1.442	1.755	.9132	.5699	2.324	5.061	5.159	.7296	.5995	7.309
.1530	.1858	1.167	.8234	1.451	1.762	.9132	.5675	2.335	5.115	5.212	.7282	.5996	7.279
.1540	.1866	1.173	.8250	1.460	1.770	.9132	.5651	2.345	5.169	5.265	.7268	.5996	7.250
.1550	.1875	1.178	.8267	1.469	1.777	.9131	.5627	2.356	5.225	5.320	.7254	.5997	7.221
.1560	.1883	1.183	.8284	1.479	1.785	.9130	.5602	2.366	5.283	5.376	.7240	.5998	7.191
.1570	.1891	1.188	.8301	1.488	1.793	.9129	.5577	2.377	5.339	5.432	.7226	.5999	7.162
.1580	.1900	1.194	.8317	1.498	1.801	.9130	.5552	2.387	5.398	5.490	.7212	.5998	7.134
.1590	.1908	1.199	.8333	1.507	1.809	.9130	.5528	2.398	5.454	5.544	.7198	.5998	7.107
.1600	.1917	1.204	.8349	1.517	1.817	.9130	.5504	2.408	5.513	5.603	.7184	.5998	7.079
.1610	.1925	1.209	.8365	1.527	1.825	.9130	.5480	2.419	5.571	5.660	.7171	.5998	7.052
.1620	.1933	1.215	.8381	1.536	1.833	.9130	.5456	2.429	5.630	5.718	.7157	.5998	7.026
.1630	.1941	1.220	.8396	1.546	1.841	.9130	.5432	2.440	5.690	5.777	.7144	.5998	7.000
.1640	.1950	1.225	.8411	1.555	1.849	.9130	.5409	2.450	5.751	5.837	.7130	.5998	6.975
.1650	.1958	1.230	.8427	1.565	1.857	.9131	.5385	2.461	5.813	5.898	.7117	.5997	6.949
.1660	.1966	1.235	.8442	1.574	1.865	.9132	.5362	2.471	5.874	5.959	.7103	.5996	6.924
.1670	.1975	1.240	.8457	1.584	1.873	.9132	.5339	2.482	5.938	6.021	.7090	.5996	6.900
.1680	.1983	1.246	.8472	1.594	1.882	.9133	.5315	2.492	6.003	6.085	.7076	.5995	6.876
.1690	.1992	1.251	.8486	1.604	1.890	.9133	.5291	2.503	6.066	6.148	.7063	.5994	6.853
.1700	.2000	1.257	.8501	1.614	1.899	.9134	.5267	2.513	6.130	6.212	.7050	.5993	6.830
.1710	.2008	1.262	.8515	1.624	1.907	.9135	.5243	2.523	6.197	6.275	.7036	.5992	6.807
.1720	.2017	1.267	.8529	1.634	1.915	.9136	.5220	2.534	6.262	6.342	.7023	.5991	6.784
.1730	.2025	1.272	.8544	1.644	1.924	.9137	.5197	2.544	6.329	6.407	.7010	.5989	6.761
.1740	.2033	1.277	.8558	1.654	1.933	.9138	.5174	2.555	6.395	6.473	.6997	.5988	6.738
.1750	.2042	1.282	.8572	1.664	1.941	.9139	.5151	2.565	6.465	6.541	.6984	.5987	6.716
.1760	.2050	1.288	.8586	1.675	1.951	.9140	.5127	2.576	6.534	6.610	.6971	.5985	6.694
.1770	.2058	1.293	.8600	1.685	1.959	.9141	.5104	2.586	6.603	6.679	.6958	.5984	6.672
.1780	.2066	1.298	.8614	1.695	1.968	.9142	.5081	2.597	6.672	6.747	.6946	.5982	6.651
.1790	.2075	1.304	.8627	1.706	1.977	.9144	.5058	2.607	6.744	6.818	.6933	.5980	6.631
.1800	.2083	1.309	.8640	1.716	1.986	.9145	.5036	2.618	6.818	6.891	.6920	.5979	6.611
.1810	.2092	1.314	.8653	1.727	1.995	.9146	.5013	2.629	6.890	6.963	.6907	.5977	6.591
.1820	.2100	1.320	.8666	1.737	2.004	.9148	.4990	2.639	6.963	7.035	.6895	.5975	6.571
.1830	.2108	1.325	.8680	1.748	2.013	.9149	.4967	2.650	7.038	7.109	.6882	.5974	6.550
.1840	.2117	1.330	.8693	1.758	2.022	.9150	.4945	2.660	7.113	7.183	.6870	.5972	6.530
.1850	.2125	1.335	.8706	1.769	2.032	.9152	.4922	2.671	7.191	7.260	.6857	.5969	6.511
.1860	.2134	1.341	.8718	1.780	2.041	.9154	.4899	2.681	7.267	7.336	.6845	.5967	6.492
.1870	.2142	1.346	.8731	1.791	2.051	.9155	.4876	2.692	7.345	7.412	.6832	.5965	6.474
.1880	.2150	1.351	.8743	1.801	2.060	.9157	.4854	2.702	7.421	7.488	.6820	.5963	6.456
.1890	.2159	1.356	.8755	1.812	2.070	.9159	.4832	2.712	7.500	7.566	.6808	.5961	6.438
.1900	.2167	1.362	.8767	1.823	2.079	.9161	.4809	2.723	7.581	7.647	.6796	.5958	6.421
.1910	.2176	1.367	.8779	1.834	2.089	.9163	.4787	2.734	7.663	7.728	.6784	.5955	6.403
.1920	.2184	1.372	.8791	1.845	2.099	.9165	.4765	2.744	7.746	7.810	.6772	.5952	6.385
.1930	.2192	1.377	.8803	1.856	2.108	.9167	.4743	2.755	7.827	7.891	.6760	.5950	6.368
.1940	.2201	1.383	.8815	1.867	2.118	.9169	.4721	2.765	7.911	7.974	.6748	.5948	6.351
.1950	.2209	1.388	.8827	1.879	2.128	.9170	.4699	2.776	7.996	8.059	.6736	.5946	6.334
.1960	.2218	1.393	.8839	1.890	2.138	.9172	.4677	2.787	8.083	8.145	.6724	.5944	6.317
.1970	.2226	1.399	.8850	1.901	2.148	.9174	.4655	2.797	8.167	8.228	.6712	.5941	6.300
.1980	.2234	1.404	.8862	1.913	2.158	.9176	.4633	2.808	8.256	8.316	.6700	.5938	6.284
.1990	.2243	1.409	.8873	1.924	2.169	.9179	.4611	2.819	8.346	8.406	.6689	.5935	6.268
.2000	.2251	1.414	.8884	1.935	2.178	.9181	.4590	2.829	8.436	8.495	.6677	.5932	6.253
.2010	.2260	1.420	.8895	1.947	2.189	.9183	.4569	2.840	8.524	8.583	.6666	.5929	6.237
.2020	.2268	1.425	.8906	1.959	2.199	.9186	.4547	2.850	8.616	8.674	.6654	.5926	6.222
.2030	.2277	1.430	.8917	1.970	2.210	.9188	.4526	2.861	8.708	8.766	.6642	.5923	6.206
.2040	.2285	1.436	.8928	1.982	2.220	.9190	.4504	2.872	8.803	8.860	.6631	.5920	6.191
.2050	.2293	1.441	.8939	1.994	2.231	.9193	.4483	2.882	8.897	8.953	.6620	.5917	6.176
.2060	.2302	1.446	.8950	2.006	2.242	.9195	.4462	2.893	8.994	9.050	.6608	.5914	6.161
.2070	.2310	1.451	.8960	2.017	2.252	.9197	.4441	2.903	9.090	9.144	.6597	.5911	6.147
.2080	.2319	1.457	.8971	2.030	2.263	.9200	.4419	2.914	9.187	9.240	.6586	.5908	6.133
.2090	.2328	1.462	.8981	2.042	2.274	.9202	.4398	2.925	9.288	9.342	.6574	.5905	6.119

Table C–1 – Continued

d/L_0	d/L	$2\pi d/L$	TANH $2\pi d/L$	SINH $2\pi d/L$	COSH $2\pi d/L$	H/H'_0	K	$4\pi d/L$	SINH $4\pi d/L$	COSH $4\pi d/L$	n	c_G/c_0	M
.2100	.2336	1.468	.8991	2.055	2.285	.9205	.4377	2.936	9.389	9.442	.6563	.5901	6.105
.2110	.2344	1.473	.9001	2.066	2.295	.9207	.4357	2.946	9.490	9.542	.6552	.5898	6.091
.2120	.2353	1.479	.9011	2.079	2.307	.9210	.4336	2.957	9.590	9.642	.6541	.5894	6.077
.2130	.2361	1.484	.9021	2.091	2.318	.9213	.4315	2.967	9.693	9.744	.6531	.5891	6.064
.2140	.2370	1.489	.9031	2.103	2.329	.9215	.4294	2.978	9.796	9.847	.6520	.5888	6.051
.2150	.2378	1.494	.9041	2.115	2.340	.9218	.4274	2.989	9.902	9.952	.6509	.5884	6.037
.2160	.2387	1.500	.9051	2.128	2.351	.9221	.4253	2.999	10.01	10.06	.6498	.5881	6.024
.2170	.2395	1.506	.9061	2.142	2.364	.9223	.4232	3.010	10.12	10.17	.6488	.5878	6.011
.2180	.2404	1.511	.9070	2.154	2.375	.9226	.4211	3.021	10.23	10.28	.6477	.5874	5.999
.2190	.2412	1.516	.9079	2.166	2.386	.9228	.4191	3.031	10.34	10.38	.6467	.5871	5.987
.2200	.2421	1.521	.9088	2.178	2.397	.9231	.4171	3.042	10.45	10.50	.6456	.5868	5.975
.2210	.2429	1.526	.9097	2.192	2.409	.9234	.4151	3.052	10.56	10.61	.6446	.5864	5.963
.2220	.2438	1.532	.9107	2.204	2.421	.9236	.4131	3.063	10.68	10.72	.6436	.5861	5.951
.2230	.2446	1.537	.9116	2.218	2.433	.9239	.4111	3.074	10.79	10.84	.6425	.5857	5.939
.2240	.2455	1.542	.9125	2.230	2.444	.9242	.4091	3.085	10.91	10.95	.6414	.5854	5.927
.2250	.2463	1.548	.9134	2.244	2.457	.9245	.4071	3.095	11.02	11.07	.6404	.5850	5.915
.2260	.2472	1.553	.9143	2.257	2.469	.9248	.4051	3.106	11.15	11.19	.6394	.5846	5.903
.2270	.2481	1.559	.9152	2.271	2.481	.9251	.4031	3.117	11.27	11.31	.6383	.5842	5.891
.2280	.2489	1.564	.9161	2.284	2.493	.9254	.4011	3.128	11.39	11.44	.6373	.5838	5.880
.2290	.2498	1.569	.9170	2.297	2.506	.9258	.3991	3.138	11.51	11.56	.6363	.5834	5.869
.2300	.2506	1.575	.9178	2.311	2.518	.9261	.3971	3.149	11.64	11.68	.6353	.5830	5.858
.2310	.2515	1.580	.9186	2.325	2.531	.9264	.3952	3.160	11.77	11.81	.6343	.5826	5.848
.2320	.2523	1.585	.9194	2.338	2.543	.9267	.3932	3.171	11.90	11.93	.6333	.5823	5.838
.2330	.2532	1.591	.9203	2.352	2.556	.9270	.3912	3.182	12.03	12.07	.6323	.5819	5.827
.2340	.2540	1.596	.9211	2.366	2.569	.9273	.3893	3.192	12.15	12.19	.6313	.5815	5.816
.2350	.2549	1.602	.9219	2.380	2.581	.9276	.3874	3.203	12.29	12.33	.6304	.5811	5.806
.2360	.2558	1.607	.9227	2.393	2.594	.9279	.3855	3.214	12.43	12.47	.6294	.5807	5.796
.2370	.2566	1.612	.9235	2.408	2.607	.9282	.3836	3.225	12.55	12.59	.6284	.5804	5.786
.2380	.2575	1.618	.9243	2.422	2.620	.9285	.3816	3.236	12.69	12.73	.6275	.5800	5.776
.2390	.2584	1.623	.9251	2.436	2.634	.9288	.3797	3.247	12.83	12.87	.6265	.5796	5.766
.2400	.2592	1.629	.9259	2.450	2.647	.9291	.3779	3.257	12.97	13.01	.6256	.5792	5.756
.2410	.2601	1.634	.9267	2.464	2.660	.9294	.3760	3.268	13.11	13.15	.6246	.5788	5.746
.2420	.2610	1.640	.9275	2.480	2.674	.9298	.3741	3.279	13.26	13.30	.6237	.5784	5.736
.2430	.2618	1.645	.9282	2.494	2.687	.9301	.3722	3.290	13.40	13.44	.6228	.5780	5.727
.2440	.2627	1.650	.9289	2.508	2.700	.9304	.3704	3.301	13.55	13.59	.6218	.5776	5.718
.2450	.2635	1.656	.9296	2.523	2.714	.9307	.3685	3.312	13.70	13.73	.6209	.5272	5.710
.2460	.2644	1.661	.9304	2.538	2.728	.9310	.3666	3.323	13.85	13.88	.6200	.5768	5.701
.2470	.2653	1.667	.9311	2.553	2.742	.9314	.3648	3.334	14.00	14.04	.6191	.5764	5.692
.2480	.2661	1.672	.9318	2.568	2.755	.9317	.3629	3.344	14.15	14.19	.6182	.5760	5.684
.2490	.2670	1.678	.9325	2.583	2.770	.9320	.3610	3.355	14.31	14.35	.6173	.5756	5.675
.2500	.2679	1.683	.9332	2.599	2.784	.9323	.3592	3.367	14.47	14.51	.6164	.5752	5.667
.2510	.2687	1.689	.9339	2.614	2.798	.9327	.3574	3.377	14.62	14.66	.6155	.5748	5.658
.2520	.2696	1.694	.9346	2.629	2.813	.9330	.3556	3.388	14.79	14.82	.6146	.5744	5.650
.2530	.2705	1.700	.9353	2.645	2.828	.9333	.3537	3.399	14.95	14.99	.6137	.5740	5.641
.2540	.2714	1.705	.9360	2.660	2.842	.9336	.3519	3.410	15.12	15.15	.6128	.5736	5.633
.2550	.2722	1.711	.9367	2.676	2.856	.9340	.3501	3.421	15.29	15.32	.6120	.5732	5.624
.2560	.2731	1.716	.9374	2.691	2.871	.9343	.3483	3.432	15.45	15.49	.6111	.5728	5.616
.2570	.2740	1.722	.9381	2.707	2.886	.9346	.3465	3.443	15.63	15.66	.6102	.5724	5.608
.2580	.2749	1.727	.9388	2.723	2.901	.9349	.3447	3.454	15.80	15.83	.6093	.5720	5.600
.2590	.2757	1.732	.9394	2.739	2.916	.9353	.3430	3.465	15.97	16.00	.6085	.5716	5.592
.2600	.2766	1.738	.9400	2.755	2.931	.9356	.3412	3.476	16.15	16.18	.6076	.5712	5.585
.2610	.2775	1.744	.9406	2.772	2.946	.9360	.3394	3.487	16.33	16.36	.6068	.5707	5.578
.2620	.2784	1.749	.9412	2.788	2.962	.9363	.3376	3.498	16.51	16.54	.6060	.5703	5.571
.2630	.2792	1.755	.9418	2.804	2.977	.9367	.3359	3.509	16.69	16.73	.6052	.5699	5.563
.2640	.2801	1.760	.9425	2.820	2.992	.9370	.3342	3.520	16.88	16.91	.6043	.5695	5.556
.2650	.2810	1.766	.9431	2.837	3.008	.9373	.3325	3.531	17.07	17.10	.6035	.5691	5.548
.2660	.2819	1.771	.9437	2.853	3.023	.9377	.3308	3.542	17.26	17.28	.6027	.5687	5.541
.2670	.2827	1.776	.9443	2.870	3.039	.9380	.3291	3.553	17.45	17.45	.6018	.5683	5.534
.2680	.2836	1.782	.9449	2.886	3.055	.9383	.3274	3.564	17.64	17.67	.6010	.5679	5.527
.2690	.2845	1.788	.9455	2.904	3.071	.9386	.3256	3.575	17.84	17.87	.6002	.5675	5.520

Table C–1 — Continued

d/L_o	d/L	$2\pi d/L$	TANH $2\pi d/L$	SINH $2\pi d/L$	COSH $2\pi d/L$	H/H'_o	K	$4\pi d/L$	SINH $4\pi d/L$	COSH $4\pi d/L$	n	C_G/C_o	M
.2700	.2854	1.793	.9461	2.921	3.088	.9390	.3239	3.587	18.04	18.07	.5994	.5671	5.513
.2710	.2863	1.799	.9467	2.938	3.104	.9393	.3222	3.598	18.24	18.27	.5986	.5667	5.506
.2720	.2872	1.804	.9473	2.956	3.120	.9396	.3205	3.610	18.46	18.49	.5978	.5663	5.499
.2730	.2880	1.810	.9478	2.973	3.136	.9400	.3189	3.620	18.65	18.67	.5971	.5659	5.493
.2740	.2889	1.815	.9484	2.990	3.153	.9403	.3172	3.631	18.86	18.89	.5963	.5655	5.486
.2750	.2898	1.821	.9490	3.008	3.170	.9406	.3155	3.642	19.07	19.10	.5955	.5651	5.480
.2760	.2907	1.826	.9495	3.025	3.186	.9410	.3139	3.653	19.28	19.30	.5947	.5647	5.474
.2770	.2916	1.832	.9500	3.043	3.203	.9413	.3122	3.664	19.49	19.51	.5940	.5643	5.468
.2780	.2924	1.837	.9505	3.061	3.220	.9416	.3106	3.675	19.71	19.74	.5932	.5639	5.462
.2790	.2933	1.843	.9511	3.079	3.237	.9420	.3089	3.686	19.93	19.96	.5925	.5635	5.456
.2800	.2942	1.849	.9516	3.097	3.254	.9423	.3073	3.697	20.16	20.18	.5917	.5631	5.450
.2810	.2951	1.854	.9521	3.115	3.272	.9426	.3057	3.709	20.39	20.41	.5910	.5627	5.444
.2820	.2960	1.860	.9526	3.133	3.289	.9430	.3040	3.720	20.62	20.64	.5902	.5623	5.438
.2830	.2969	1.866	.9532	3.152	3.307	.9433	.3024	3.731	20.85	20.87	.5895	.5619	5.432
.2840	.2978	1.871	.9537	3.171	3.325	.9436	.3008	3.742	21.09	21.11	.5887	.5615	5.426
.2850	.2987	1.877	.9542	3.190	3.343	.9440	.2992	3.754	21.33	21.35	.5880	.5611	5.420
.2860	.2996	1.882	.9547	3.209	3.361	.9443	.2976	3.765	21.57	21.59	.5873	.5607	5.414
.2870	.3005	1.888	.9552	3.228	3.379	.9446	.2959	3.776	21.82	21.84	.5866	.5603	5.409
.2880	.3014	1.893	.9557	3.246	3.396	.9449	.2944	3.787	22.05	22.07	.5859	.5600	5.403
.2890	.3022	1.899	.9562	3.264	3.414	.9452	.2929	3.798	22.30	22.32	.5852	.5596	5.397
.2900	.3031	1.905	.9567	3.284	3.433	.9456	.2913	3.809	22.54	22.57	.5845	.5592	5.392
.2910	.3040	1.910	.9572	3.303	3.451	.9459	.2898	3.821	22.81	22.83	.5838	.5588	5.386
.2920	.3049	1.916	.9577	3.323	3.471	.9463	.2882	3.832	23.07	23.09	.5831	.5584	5.380
.2930	.3058	1.922	.9581	3.343	3.490	.9466	.2866	3.843	23.33	23.35	.5824	.5580	5.375
.2940	.3067	1.927	.9585	3.362	3.508	.9469	.2851	3.855	23.60	23.62	.5817	.5576	5.371
.2950	.3076	1.933	.9590	3.382	3.527	.9473	.2835	3.866	23.86	23.88	.5810	.5572	5.366
.2960	.3085	1.938	.9594	3.402	3.546	.9476	.2820	3.877	24.12	24.15	.5804	.5568	5.361
.2970	.3094	1.944	.9599	3.422	3.565	.9480	.2805	3.888	24.40	24.42	.5797	.5564	5.356
.2980	.3103	1.950	.9603	3.442	3.585	.9483	.2790	3.900	24.68	24.70	.5790	.5560	5.351
.2990	.3112	1.955	.9607	3.462	3.604	.9486	.2775	3.911	24.96	24.98	.5784	.5556	5.347
.3000	.3121	1.961	.9611	3.483	3.624	.9490	.2760	3.922	25.24	25.26	.5777	.5552	5.342
.3010	.3130	1.967	.9616	3.503	3.643	.9493	.2745	3.933	25.53	25.55	.5771	.5549	5.337
.3020	.3139	1.972	.9620	3.524	3.663	.9496	.2730	3.945	25.82	25.83	.5764	.5545	5.332
.3030	.3148	1.978	.9624	3.545	3.683	.9499	.2715	3.956	26.12	26.14	.5758	.5541	5.328
.3040	.3157	1.984	.9629	3.566	3.703	.9502	.2700	3.968	26.42	26.44	.5751	.5538	5.323
.3050	.3166	1.989	.9633	3.587	3.724	.9505	.2685	3.979	26.72	26.74	.5745	.5534	5.318
.3060	.3175	1.995	.9637	3.609	3.745	.9509	.2670	3.990	27.02	27.04	.5739	.5530	5.314
.3070	.3184	2.001	.9641	3.630	3.765	.9512	.2656	4.002	27.33	27.35	.5732	.5527	5.309
.3080	.3193	2.007	.9645	3.651	3.786	.9515	.2641	4.013	27.65	27.66	.5726	.5523	5.305
.3090	.3202	2.012	.9649	3.673	3.806	.9518	.2627	4.024	27.96	27.98	.5720	.5519	5.300
.3100	.3211	2.018	.9653	3.694	3.827	.9522	.2613	4.036	28.28	28.30	.5714	.5515	5.296
.3110	.3220	2.023	.9656	3.716	3.848	.9525	.2599	4.047	28.60	28.62	.5708	.5511	5.292
.3120	.3230	2.029	.9660	3.738	3.870	.9528	.2584	4.058	28.93	28.95	.5701	.5507	5.288
.3130	.3239	2.035	.9664	3.760	3.891	.9531	.2570	4.070	29.27	29.28	.5695	.5504	5.284
.3140	.3248	2.041	.9668	3.782	3.912	.9535	.2556	4.081	29.60	29.62	.5689	.5500	5.280
.3150	.3257	2.046	.9672	3.805	3.934	.9538	.2542	4.093	29.94	29.96	.5683	.5497	5.276
.3160	.3266	2.052	.9676	3.828	3.956	.9541	.2528	4.104	30.29	30.31	.5678	.5494	5.272
.3170	.3275	2.058	.9679	3.851	3.978	.9544	.2514	4.116	30.64	30.65	.5672	.5490	5.268
.3180	.3284	2.063	.9682	3.873	4.000	.9547	.2500	4.127	30.99	31.00	.5666	.5486	5.264
.3190	.3294	2.069	.9686	3.896	4.022	.9550	.2486	4.139	31.35	31.37	.5660	.5483	5.260
.3200	.3302	2.075	.9690	3.919	4.045	.9553	.2472	4.150	31.71	31.72	.5655	.5479	5.256
.3210	.3311	2.081	.9693	3.943	4.068	.9556	.2459	4.161	32.07	32.08	.5649	.5476	5.252
.3220	.3321	2.086	.9696	3.966	4.090	.9559	.2445	4.173	32.44	32.46	.5643	.5472	5.249
.3230	.3330	2.092	.9700	3.990	4.114	.9562	.2431	4.185	32.83	32.84	.5637	.5468	5.245
.3240	.3339	2.098	.9703	4.014	4.136	.9565	.2418	4.196	33.20	33.22	.5632	.5465	5.241
.3250	.3349	2.104	.9707	4.038	4.160	.9568	.2404	4.208	33.60	33.61	.5627	.5462	5.237
.3260	.3357	2.110	.9710	4.061	4.183	.9571	.2391	4.219	33.97	33.99	.5621	.5458	5.234
.3270	.3367	2.115	.9713	4.085	4.206	.9574	.2378	4.231	34.37	34.38	.5616	.5455	5.231
.3280	.3376	2.121	.9717	4.110	4.230	.9577	.2364	4.242	34.77	34.79	.5610	.5451	5.227
.3290	.3385	2.127	.9720	4.135	4.254	.9580	.2351	4.254	35.18	35.19	.5605	.5448	5.223

Table C—1 — Continued

d/L_o	d/L	$2\pi d/L$	TANH $2\pi d/L$	SINH $2\pi d/L$	COSH $2\pi d/L$	H/H'_o	K	$4\pi d/L$	SINH $4\pi d/L$	COSH $4\pi d/L$	n	C_G/C_o	M
.3300	.3394	2.133	.9723	4.159	4.277	.9583	.2338	4.265	35.58	35.59	.5599	.5444	5.220
.3310	.3403	2.138	.9726	4.184	4.301	.9586	.2325	4.277	35.99	36.00	.5594	.5441	5.217
.3320	.3413	2.144	.9729	4.209	4.326	.9589	.2312	4.288	36.42	36.43	.5589	.5438	5.214
.3330	.3422	2.150	.9732	4.234	4.350	.9592	.2299	4.300	36.84	36.85	.5584	.5434	5.210
.3340	.3431	2.156	.9735	4.259	4.375	.9595	.2286	4.311	37.25	37.27	.5578	.5431	5.207
.3350	.3440	2.161	.9738	4.284	4.399	.9598	.2273	4.323	37.70	37.72	.5573	.5427	5.204
.3360	.3449	2.167	.9741	4.310	4.424	.9601	.2260	4.335	38.14	38.15	.5568	.5424	5.201
.3370	.3459	2.173	.9744	4.336	4.450	.9604	.2247	4.346	38.59	38.60	.5563	.5421	5.198
.3380	.3468	2.179	.9747	4.361	4.474	.9607	.2235	4.358	39.02	39.04	.5558	.5417	5.194
.3390	.3477	2.185	.9750	4.388	4.500	.9610	.2222	4.369	39.48	39.49	.5553	.5414	5.191
.3400	.3468	2.190	.9753	4.413	4.525	.9613	.2210	4.381	39.95	39.96	.5548	.5411	5.188
.3410	.3495	2.196	.9756	4.439	4.550	.9615	.2198	4.392	40.40	40.41	.5544	.5408	5.185
.3420	.3504	2.202	.9758	4.466	4.576	.9618	.2185	4.404	40.87	40.89	.5539	.5405	5.182
.3430	.3514	2.208	.9761	4.492	4.602	.9621	.2173	4.416	41.36	41.37	.5534	.5402	5.179
.3440	.3523	2.214	.9764	4.521	4.630	.9623	.2160	4.427	41.85	41.84	.5529	.5399	5.176
.3450	.3532	2.220	.9767	4.547	4.656	.9626	.2148	4.439	42.33	42.34	.5524	.5396	5.173
.3460	.3542	2.225	.9769	4.575	4.682	.9629	.2136	4.451	42.83	42.84	.5519	.5392	5.171
.3470	.3551	2.231	.9772	4.602	4.709	.9632	.2124	4.462	43.34	43.35	.5515	.5389	5.168
.3480	.3560	2.237	.9775	4.629	4.736	.9635	.2111	4.474	43.85	43.86	.5510	.5386	5.165
.3490	.3570	2.243	.9777	4.657	4.763	.9638	.2099	4.486	44.37	44.40	.5505	.5383	5.162
.3500	.3579	2.249	.9780	4.685	4.791	.9640	.2087	4.498	44.89	44.80	.5501	.5380	5.159
.3510	.3588	2.255	.9782	4.713	4.818	.9643	.2076	4.509	45.42	45.43	.5496	.5377	5.157
.3520	.3598	2.260	.9785	4.741	4.845	.9646	.2064	4.521	45.95	45.96	.5492	.5374	5.154
.3530	.3607	2.266	.9787	4.770	4.873	.9648	.2052	4.533	46.50	46.51	.5487	.5371	5.152
.3540	.3616	2.272	.9790	4.798	4.901	.9651	.2040	4.544	47.03	47.04	.5483	.5368	5.149
.3550	.3625	2.278	.9792	4.827	4.929	.9654	.2029	4.556	47.59	47.60	.5479	.5365	5.147
.3560	.3635	2.284	.9795	4.856	4.957	.9657	.2017	4.568	48.15	48.16	.5474	.5362	5.144
.3570	.3644	2.290	.9797	4.885	4.987	.9659	.2005	4.579	48.72	48.73	.5470	.5359	5.141
.3580	.3653	2.296	.9799	4.914	5.015	.9662	.1994	4.591	49.29	49.30	.5466	.5356	5.139
.3590	.3663	2.301	.9801	4.944	5.044	.9665	.1983	4.603	49.88	49.89	.5461	.5353	5.137
.3600	.3672	2.307	.9804	4.974	5.072	.9667	.1972	4.615	50.47	50.48	.5457	.5350	5.134
.3610	.3682	2.313	.9806	5.004	5.103	.9670	.1960	4.627	51.08	51.09	.5453	.5347	5.132
.3620	.3691	2.319	.9808	5.034	5.132	.9673	.1949	4.638	51.67	51.67	.5449	.5344	5.130
.3630	.3700	2.325	.9811	5.063	5.161	.9675	.1938	4.650	52.27	52.28	.5445	.5342	5.127
.3640	.3709	2.331	.9813	5.094	5.191	.9677	.1926	4.661	52.89	52.90	.5441	.5339	5.125
.3650	.3719	2.337	.9815	5.124	5.221	.9680	.1915	4.673	53.52	53.53	.5437	.5336	5.123
.3660	.3728	2.342	.9817	5.155	5.251	.9683	.1904	4.685	54.15	54.16	.5433	.5333	5.121
.3670	.3737	2.348	.9819	5.186	5.281	.9686	.1894	4.697	54.78	54.79	.5429	.5330	5.118
.3680	.3747	2.354	.9821	5.217	5.312	.9688	.1883	4.708	55.42	55.43	.5425	.5327	5.116
.3690	.3756	2.360	.9823	5.248	5.343	.9690	.1872	4.720	56.09	56.10	.5421	.5325	5.114
.3700	.3766	2.366	.9825	5.280	5.374	.9693	.1861	4.732	56.76	56.77	.5417	.5322	5.112
.3710	.3775	2.372	.9827	5.312	5.406	.9696	.1850	4.744	57.43	57.44	.5413	.5319	5.110
.3720	.3785	2.378	.9830	5.345	5.438	.9698	.1839	4.756	58.13	58.14	.5409	.5317	5.107
.3730	.3794	2.384	.9832	5.377	5.469	.9700	.1828	4.768	58.82	58.83	.5405	.5314	5.105
.3740	.3804	2.390	.9834	5.410	5.502	.9702	.1818	4.780	59.52	59.53	.5402	.5312	5.103
.3750	.3813	2.396	.9835	5.443	5.534	.9705	.1807	4.792	60.24	60.25	.5398	.5309	5.101
.3760	.3822	2.402	.9837	5.475	5.566	.9707	.1797	4.803	60.95	60.95	.5394	.5306	5.099
.3770	.3832	2.408	.9839	5.508	5.598	.9709	.1786	4.815	61.68	61.68	.5390	.5304	5.097
.3780	.3841	2.413	.9841	5.541	5.631	.9712	.1776	4.827	62.41	62.42	.5387	.5301	5.095
.3790	.3850	2.419	.9843	5.572	5.661	.9714	.1766	4.838	63.13	63.14	.5383	.5299	5.093
.3800	.3860	2.425	.9845	5.609	5.697	.9717	.1756	4.851	63.91	63.91	.5380	.5296	5.091
.3810	.3869	2.431	.9847	5.643	5.731	.9719	.1745	4.862	64.67	64.67	.5376	.5294	5.090
.3820	.3879	2.437	.9848	5.677	5.765	.9721	.1735	4.875	65.45	65.46	.5372	.5291	5.088
.3830	.3888	2.443	.9850	5.712	5.798	.9724	.1725	4.885	66.16	66.17	.5369	.5288	5.086
.3840	.3898	2.449	.9852	5.746	5.833	.9726	.1715	4.898	67.02	67.03	.5365	.5286	5.084
.3850	.3907	2.455	.9854	5.780	5.866	.9728	.1705	4.910	67.80	67.81	.5362	.5284	5.082
.3860	.3917	2.461	.9855	5.814	5.900	.9730	.1695	4.922	68.61	68.62	.5359	.5281	5.081
.3870	.3926	2.467	.9857	5.850	5.935	.9732	.1685	4.934	69.45	69.46	.5355	.5279	5.079
.3880	.3936	2.473	.9859	5.886	5.970	.9735	.1675	4.946	70.28	70.29	.5352	.5276	5.077
.3890	.3945	2.479	.9860	5.921	6.005	.9737	.1665	4.958	71.12	71.13	.5349	.5274	5.076

Table C–1 – Continued

d/L_o	d/L	$2\pi d/L$	TANH $2\pi d/L$	SINH $2\pi d/L$	COSH $2\pi d/L$	H/H'_o	K	$4\pi d/L$	SINH $4\pi d/L$	COSH $4\pi d/L$	n	C_G/C_o	M
.3900	.3955	2.485	.9862	5.957	6.040	.9739	.1656	4.970	71.97	71.98	.5345	.5271	5.074
.3910	.3964	2.491	.9864	5.993	6.076	.9741	.1646	4.982	72.85	72.86	.5342	.5269	5.072
.3920	.3974	2.497	.9865	6.029	6.112	.9743	.1636	4.993	73.72	73.72	.5339	.5267	5.071
.3930	.3983	2.503	.9867	6.066	6.148	.9745	.1627	5.005	74.58	74.59	.5336	.5265	5.069
.3940	.3993	2.509	.9869	6.103	6.185	.9748	.1617	5.017	75.48	75.49	.5332	.5262	5.067
.3950	.4002	2.515	.9870	6.140	6.221	.9750	.1608	5.029	76.40	76.40	.5329	.5260	5.066
.3960	.4012	2.521	.9872	6.177	6.258	.9752	.1598	5.041	77.31	77.32	.5326	.5258	5.064
.3970	.4021	2.527	.9873	6.215	6.295	.9754	.1589	5.053	78.24	78.24	.5323	.5255	5.063
.3980	.4031	2.532	.9874	6.252	6.332	.9756	.1579	5.065	79.19	79.19	.5320	.5253	5.062
.3990	.4040	2.538	.9876	6.290	6.369	.9758	.1570	5.077	80.13	80.13	.5317	.5251	5.060
.4000	.4050	2.544	.9877	6.329	6.407	.9761	.1561	5.089	81.12	81.12	.5314	.5248	5.058
.4010	.4059	2.550	.9879	6.367	6.445	.9763	.1552	5.101	82.07	82.08	.5311	.5246	5.056
.4020	.4069	2.556	.9880	6.406	6.483	.9765	.1542	5.113	83.06	83.06	.5308	.5244	5.055
.4030	.4078	2.562	.9882	6.444	6.521	.9766	.1533	5.125	84.07	84.07	.5305	.5242	5.053
.4040	.4088	2.568	.9883	6.484	6.561	.9768	.1524	5.137	85.11	85.12	.5302	.5240	5.052
.4050	.4098	2.575	.9885	6.525	6.601	.9770	.1515	5.149	86.14	86.14	.5299	.5238	5.050
.4060	.4107	2.581	.9886	6.564	6.640	.9772	.1506	5.161	87.17	87.17	.5296	.5236	5.049
.4070	.4116	2.586	.9887	6.603	6.679	.9774	.1497	5.173	88.19	88.20	.5293	.5234	5.048
.4080	.4126	2.592	.9889	6.644	6.718	.9776	.1488	5.185	89.28	89.28	.5290	.5232	5.046
.4090	.4136	2.598	.9890	6.684	6.758	.9778	.1480	5.197	90.38	90.39	.5287	.5229	5.045
.4100	.4145	2.604	.9891	6.725	6.799	.9780	.1471	5.209	91.44	91.44	.5285	.5227	5.044
.4110	.4155	2.610	.9892	6.766	6.839	.9782	.1462	5.221	92.54	92.55	.5282	.5225	5.043
.4120	.4164	2.616	.9894	6.806	6.879	.9784	.1454	5.233	93.67	93.67	.5279	.5223	5.041
.4130	.4174	2.623	.9895	6.849	6.921	.9786	.1445	5.245	94.83	94.83	.5277	.5221	5.040
.4140	.4183	2.629	.9896	6.890	6.963	.9788	.1436	5.257	95.95	95.96	.5274	.5219	5.039
.4150	.4193	2.635	.9898	6.932	7.004	.9790	.1428	5.269	97.13	97.13	.5271	.5217	5.037
.4160	.4203	2.641	.9899	6.974	7.046	.9792	.1419	5.281	98.29	98.30	.5269	.5215	5.036
.4170	.4212	2.647	.9900	7.018	7.088	.9794	.1411	5.294	99.52	99.52	.5266	.5213	5.035
.4180	.4222	2.653	.9901	7.060	7.130	.9795	.1403	5.305	100.7	100.7	.5263	.5211	5.034
.4190	.4231	2.659	.9902	7.102	7.173	.9797	.1394	5.317	101.9	101.9	.5261	.5209	5.033
.4200	.4241	2.665	.9904	7.146	7.215	.9798	.1386	5.329	103.1	103.1	.5258	.5208	5.031
.4210	.4251	2.671	.9905	7.190	7.259	.9800	.1378	5.341	104.4	104.4	.5256	.5206	5.030
.4220	.4260	2.677	.9906	7.234	7.303	.9802	.1369	5.353	105.7	105.7	.5253	.5204	5.029
.4230	.4270	2.683	.9907	7.279	7.349	.9804	.1361	5.366	107.0	107.0	.5251	.5202	5.028
.4240	.4280	2.689	.9908	7.325	7.392	.9806	.1353	5.378	108.3	108.3	.5248	.5200	5.027
.4250	.4289	2.695	.9909	7.371	7.438	.9808	.1345	5.390	109.	109.7	.5246	.5198	5.026
.4260	.4298	2.701	.9910	7.412	7.479	.9810	.1337	5.402	110.9	110.9	.5244	.5196	5.025
.4270	.4308	2.707	.9911	7.457	7.524	.9811	.1329	5.414	112.2	112.2	.5241	.5195	5.024
.4280	.4318	2.713	.9912	7.503	7.570	.9812	.1321	5.426	113.6	113.6	.5239	.5193	5.023
.4290	.4328	2.719	.9913	7.550	7.616	.9814	.1313	5.438	115.0	115.0	.5237	.5191	5.022
.4300	.4337	2.725	.9914	7.595	7.661	.9816	.1305	5.450	116.4	116.4	.5234	.5189	5.021
.4310	.4347	2.731	.9915	7.642	7.707	.9818	.1298	5.462	117.8	117.8	.5232	.5187	5.020
.4320	.4356	2.737	.9916	7.688	7.753	.9819	.1290	5.474	119.2	119.3	.5230	.5186	5.019
.4330	.4366	2.743	.9917	7.735	7.800	.9821	.1282	5.486	120.7	120.7	.5227	.5184	5.018
.4340	.4376	2.749	.9918	7.783	7.847	.9823	.1274	5.499	122.2	122.2	.5225	.5182	5.017
.4350	.4385	2.755	.9919	7.831	7.895	.9824	.1267	5.511	123.7	123.7	.5223	.5181	5.016
.4360	.4395	2.762	.9920	7.880	7.943	.9826	.1259	5.523	125.2	125.2	.5221	.5179	5.015
.4370	.4405	2.768	.9921	7.922	7.991	.9828	.1251	5.535	126.7	126.7	.5218	.5177	5.014
.4380	.4414	2.774	.9922	7.975	8.035	.9829	.1244	5.547	128.3	128.3	.5216	.5176	5.013
.4390	.4424	2.780	.9923	8.026	8.088	.9830	.1236	5.560	129.9	129.9	.5214	.5174	5.012
.4400	.4434	2.786	.9924	8.075	8.136	.9832	.1229	5.572	131.4	131.4	.5212	.5172	5.011
.4410	.4443	2.792	.9925	8.124	8.185	.9833	.1222	5.584	133.0	133.0	.5210	.5171	5.010
.4420	.4453	2.798	.9926	8.175	8.236	.9835	.1214	5.596	134.7	134.7	.5208	.5169	5.009
.4430	.4463	2.804	.9927	8.228	8.285	.9836	.1207	5.608	136.3	136.3	.5206	.5168	5.008
.4440	.4472	2.810	.9928	8.274	8.334	.9838	.1200	5.620	137.9	137.9	.5204	.5166	5.007
.4450	.4482	2.816	.9929	8.326	8.387	.9839	.1192	5.632	139.6	139.7	.5202	.5165	5.006
.4460	.4492	2.822	.9930	8.379	8.438	.9841	.1185	5.644	141.4	141.4	.5200	.5163	5.005
.4470	.4501	2.828	.9930	8.427	8.486	.9843	.1178	5.657	143.1	143.1	.5198	.5161	5.005
.4480	.4511	2.834	.9931	8.481	8.540	.9844	.1171	5.669	144.8	144.8	.5196	.5160	5.004
.4490	.4521	2.840	.9932	8.532	8.590	.9846	.1164	5.681	146.6	146.6	.5194	.5158	5.003

Table C–1 — Continued

d/L_o	d/L	$2\pi\, d/L$	TANH $2\pi\, d/L$	SINH $2\pi\, d/L$	COSH $2\pi\, d/L$	H/H'_o	K	$4\pi\, d/L$	SINH $4\pi\, d/L$	COSH $4\pi\, d/L$	n	C_g/C_o	M
.4500	.4531	2.847	.9933	8.585	8.643	.9847	.1157	5.693	148.4	148.4	.5192	.5157	5.002
.4510	.4540	2.853	.9934	8.638	8.695	.9848	.1150	5.705	150.2	150.2	.5190	.5156	5.001
.4520	.4550	2.859	.9935	8.693	8.750	.9849	.1143	5.717	152.1	152.1	.5188	.5154	5.000
.4530	.4560	2.865	.9935	8.747	8.804	.9851	.1136	5.730	154.0	154.0	.5186	.5152	5.000
.4540	.4569	2.871	.9936	8.797	8.854	.9852	.1129	5.742	155.9	155.9	.5184	.5151	4.999
.4550	.4579	2.877	.9937	8.853	8.910	.9853	.1122	5.754	157.7	157.7	.5182	.5150	4.998
.4560	.4589	2.883	.9938	8.910	8.965	.9855	.1115	5.766	159.7	159.7	.5181	.5148	4.997
.4570	.4599	2.890	.9938	8.965	9.021	.9857	.1109	5.779	161.7	161.7	.5179	.5146	4.997
.4580	.4608	2.896	.9939	9.016	9.072	.9858	.1102	5.791	163.6	163.6	.5177	.5145	4.996
.4590	.4618	2.902	.9940	9.074	9.129	.9859	.1095	5.803	165.6	165.6	.5175	.5144	4.995
.4600	.4628	2.908	.9941	9.132	9.186	.9860	.1089	5.815	167.7	167.7	.5173	.5143	4.994
.4610	.4637	2.914	.9941	9.183	9.238	.9862	.1083	5.827	169.7	169.7	.5172	.5141	4.994
.4620	.4647	2.920	.9942	9.242	9.296	.9863	.1076	5.840	171.8	171.8	.5170	.5140	4.993
.4630	.4657	2.926	.9943	9.301	9.354	.9864	.1069	5.852	173.9	173.9	.5168	.5139	4.992
.4640	.4666	2.932	.9944	9.353	9.406	.9865	.1063	5.864	176.0	176.0	.5167	.5138	4.991
.4650	.4676	2.938	.9944	9.413	9.466	.9867	.1056	5.876	178.2	178.2	.5165	.5136	4.991
.4660	.4686	2.944	.9945	9.472	9.525	.9868	.1050	5.888	180.4	180.4	.5163	.5135	4.990
.4670	.4695	2.951	.9946	9.533	9.585	.9869	.1043	5.900	182.6	182.6	.5162	.5134	4.989
.4680	.4705	2.957	.9946	9.586	9.638	.9871	.1037	5.912	184.8	184.8	.5160	.5132	4.989
.4690	.4715	2.963	.9947	9.647	9.699	.9872	.1031	5.925	187.2	187.2	.5158	.5131	4.988
.4700	.4725	2.969	.9947	9.709	9.760	.9873	.1025	5.937	189.5	189.5	.5157	.5129	4.988
.4710	.4735	2.975	.9948	9.770	9.821	.9874	.1018	5.949	191.8	191.8	.5155	.5128	4.987
.4720	.4744	2.981	.9949	9.826	9.877	.9875	.1012	5.962	194.2	194.2	.5154	.5127	4.986
.4730	.4754	2.987	.9949	9.888	9.938	.9876	.1006	5.974	196.5	196.5	.5152	.5126	4.986
.4740	.4764	2.993	.9950	9.951	10.00	.9877	.1000	5.986	199.0	199.0	.5150	.5125	4.985
.4750	.4774	2.999	.9951	10.01	10.07	.9878	.09942	5.999	201.4	201.4	.5149	.5124	4.984
.4760	.4783	3.005	.9951	10.07	10.12	.9880	.09882	6.011	203.9	203.9	.5147	.5122	4.984
.4770	.4793	3.012	.9952	10.13	10.18	.9881	.09820	6.023	206.5	206.5	.5146	.5121	4.983
.4780	.4803	3.018	.9952	10.20	10.25	.9882	.09759	6.036	209.0	209.0	.5144	.5120	4.983
.4790	.4813	3.024	.9953	10.26	10.31	.9883	.09698	6.048	211.7	211.7	.5143	.5119	4.982
.4800	.4822	3.030	.9953	10.32	10.37	.9885	.09641	6.060	214.2	214.2	.5142	.5117	4.982
.4810	.4832	3.036	.9954	10.39	10.43	.9886	.09583	6.072	216.8	216.8	.5140	.5116	4.981
.4820	.4842	3.042	.9955	10.45	10.50	.9887	.09523	6.085	219.5	219.5	.5139	.5115	4.980
.4830	.4852	3.049	.9955	10.52	10.57	.9888	.09464	6.097	222.2	222.2	.5137	.5114	4.980
.4840	.4862	3.055	.9956	10.59	10.63	.9889	.09405	6.109	225.0	225.0	.5136	.5113	4.979
.4850	.4871	3.061	.9956	10.65	10.69	.9890	.09352	6.121	228.3	228.3	.5134	.5112	4.979
.4860	.4881	3.067	.9957	10.71	10.76	.9891	.09294	6.134	230.6	230.6	.5133	.5111	4.978
.4870	.4891	3.073	.9957	10.78	10.83	.9892	.09236	6.146	233.5	233.5	.5132	.5110	4.978
.4880	.4901	3.079	.9958	10.85	10.90	.9893	.09178	6.159	236.4	236.4	.5130	.5109	4.977
.4890	.4911	3.086	.9958	10.92	10.96	.9895	.09121	6.171	239.6	239.6	.5129	.5107	4.977
.4900	.4920	3.092	.9959	10.99	11.03	.9896	.09064	6.183	242.3	242.3	.5128	.5106	4.976
.4910	.4930	3.098	.9959	11.05	11.09	.9897	.09010	6.195	245.2	245.2	.5126	.5105	4.976
.4920	.4940	3.104	.9960	11.12	11.16	.9898	.08956	6.208	248.3	248.3	.5125	.5104	4.975
.4930	.4950	3.110	.9960	11.19	11.24	.9899	.08901	6.220	251.3	251.3	.5124	.5103	4.975
.4940	.4960	3.117	.9961	11.26	11.31	.9899	.08845	6.232	254.5	254.5	.5122	.5102	4.974
.4950	.4969	3.122	.9961	11.32	11.37	.9900	.08793	6.245	257.6	257.6	.5121	.5101	4.974
.4960	.4979	3.128	.9962	11.40	11.44	.9901	.08741	6.257	260.8	260.8	.5120	.5100	4.973
.4970	.4989	3.135	.9962	11.47	11.51	.9902	.08691	6.269	264.0	264.0	.5119	.5099	4.973
.4980	.4999	3.141	.9963	11.54	11.59	.9903	.08637	6.282	267.3	267.3	.5118	.5098	4.972
.4990	.5009	3.147	.9963	11.61	11.65	.9904	.08584	6.294	270.6	270.6	.5116	.5097	4.972
.5000	.5018	3.153	.9964	11.68	11.72	.9905	.08530	6.306	274.0	274.0	.5115	.5096	4.971
.5010	.5028	3.159	.9964	11.75	11.80	.9906	.08477	6.319	277.5	277.5	.5114	.5095	4.971
.5020	.5038	3.166	.9964	11.83	11.87	.9907	.08424	6.331	280.8	280.8	.5113	.5094	4.971
.5030	.5048	3.172	.9965	11.91	11.95	.9908	.08371	6.343	284.3	284.3	.5112	.5093	4.970
.5040	.5058	3.178	.9965	11.98	12.02	.9909	.08320	6.356	287.9	287.9	.5110	.5092	4.970
.5050	.5067	3.184	.9966	12.05	12.09	.9909	.08270	6.368	291.4	291.4	.5109	.5092	4.969
.5060	.5077	3.190	.9966	12.12	12.16	.9910	.08220	6.380	295.0	295.0	.5108	.5091	4.969
.5070	.5087	3.196	.9967	12.20	12.24	.9911	.08169	6.393	298.7	298.7	.5107	.5090	4.968
.5080	.5097	3.203	.9967	12.28	12.32	.9912	.08119	6.405	302.4	302.4	.5106	.5089	4.968
.5090	.5107	3.209	.9968	12.35	12.39	.9913	.08068	6.417	306.2	306.2	.5105	.5088	4.967

Table C–1 – Continued

d/L_o	d/L	$2\pi d/L$	TANH $2\pi d/L$	SINH $2\pi d/L$	COSH $2\pi d/L$	H/H'_o	K	$4\pi d/L$	SINH $4\pi d/L$	COSH $4\pi d/L$	n	c_G/c_o	M
.5100	.5117	3.215	.9968	12.43	12.47	.9914	.08022	6.430	310.0	310.0	.5104	.5087	4.967
.5110	.5126	3.221	.9968	12.50	12.54	.9915	.07972	6.442	313.8	313.8	.5103	.5086	4.967
.5120	.5136	3.227	.9969	12.58	12.62	.9915	.07922	6.454	317.7	317.7	.5102	.5086	4.966
.5130	.5146	3.233	.9969	12.66	12.70	.9916	.07873	6.467	321.7	321.7	.5101	.5085	4.966
.5140	.5156	3.240	.9970	12.74	12.78	.9917	.07824	6.479	325.7	325.7	.5100	.5084	4.965
.5150	.5166	3.246	.9970	12.82	12.86	.9918	.07776	6.491	329.7	329.7	.5098	.5083	4.965
.5160	.5176	3.252	.9970	12.90	12.94	.9919	.07729	6.504	333.8	333.8	.5097	.5082	4.965
.5170	.5185	3.258	.9971	12.98	13.02	.9919	.07682	6.516	337.9	337.9	.5096	.5082	4.964
.5180	.5195	3.264	.9971	13.06	13.10	.9920	.07634	6.529	342.2	342.2	.5095	.5081	4.964
.5190	.5205	3.270	.9971	13.14	13.18	.9921	.07587	6.541	346.4	346.4	.5094	.5080	4.964
.5200	.5215	3.277	.9972	13.22	13.26	.9922	.07540	6.553	350.7	350.7	.5093	.5079	4.963
.5210	.5225	3.283	.9972	13.31	13.35	.9923	.07494	6.566	355.1	355.1	.5092	.5078	4.963
.5220	.5235	3.289	.9972	13.39	13.43	.9924	.07449	6.578	359.6	359.6	.5092	.5077	4.963
.5230	.5244	3.295	.9973	13.47	13.51	.9924	.07404	6.590	364.0	364.0	.5091	.5077	4.962
.5240	.5254	3.301	.9973	13.55	13.59	.9925	.07358	6.603	368.5	368.5	.5090	.5076	4.962
.5250	.5264	3.308	.9973	13.64	13.68	.9926	.07312	6.615	373.1	373.1	.5089	.5075	4.962
.5260	.5274	3.314	.9974	13.73	13.76	.9927	.07266	6.628	377.8	377.8	.5088	.5074	4.961
.5270	.5284	3.320	.9974	13.81	13.85	.9927	.07221	6.640	382.5	382.5	.5087	.5074	4.961
.5280	.5294	3.326	.9974	13.90	13.94	.9928	.07177	6.652	387.3	387.3	.5086	.5073	4.961
.5290	.5304	3.333	.9975	13.99	14.02	.9929	.07134	6.665	392.2	392.2	.5085	.5072	4.960
.5300	.5314	3.339	.9975	14.07	14.10	.9930	.07091	6.677	397.0	397.0	.5084	.5071	4.960
.5310	.5323	3.345	.9975	14.16	14.19	.9931	.07047	6.690	402.0	402.0	.5083	.5070	4.960
.5320	.5333	3.351	.9976	14.25	14.28	.9931	.07003	6.702	406.9	406.9	.5082	.5070	4.959
.5330	.5343	3.357	.9976	14.34	14.37	.9932	.06959	6.714	412.0	412.0	.5082	.5069	4.959
.5340	.5353	3.363	.9976	14.43	14.46	.9933	.06915	6.727	417.2	417.2	.5081	.5068	4.959
.5350	.5363	3.370	.9976	14.52	14.55	.9933	.06872	6.739	422.4	422.4	.5080	.5068	4.959
.5360	.5373	3.376	.9977	14.61	14.64	.9934	.06829	6.752	427.7	427.7	.5079	.5067	4.958
.5370	.5383	3.382	.9977	14.70	14.73	.9935	.06787	6.764	433.1	433.1	.5078	.5066	4.958
.5380	.5393	3.388	.9977	14.79	14.82	.9935	.06746	6.776	438.5	438.5	.5077	.5066	4.958
.5390	.5402	3.394	.9977	14.88	14.91	.9936	.06705	6.789	444.0	444.0	.5077	.5065	4.958
.5400	.5412	3.401	.9978	14.97	15.01	.9936	.06664	6.801	449.5	449.5	.5076	.5065	4.957
.5410	.5422	3.407	.9978	15.07	15.10	.9937	.06623	6.814	455.1	455.1	.5075	.5064	4.957
.5420	.5432	3.413	.9978	15.16	15.19	.9938	.06582	6.826	460.7	460.7	.5074	.5063	4.957
.5430	.5442	3.419	.9979	15.25	15.29	.9938	.06542	6.838	466.4	466.4	.5073	.5063	4.956
.5440	.5452	3.426	.9979	15.35	15.38	.9939	.06501	6.851	472.2	472.2	.5073	.5062	4.956
.5450	.5461	3.432	.9979	15.45	15.48	.9940	.06461	6.863	478.1	478.1	.5072	.5061	4.956
.5460	.5471	3.438	.9979	15.54	15.58	.9941	.06420	6.876	484.3	484.3	.5071	.5060	4.956
.5470	.5481	3.444	.9980	15.64	15.67	.9941	.06380	6.888	490.3	490.3	.5070	.5060	4.955
.5480	.5491	3.450	.9980	15.74	15.77	.9942	.06341	6.901	496.4	496.4	.5070	.5059	4.955
.5490	.5501	3.456	.9980	15.84	15.87	.9942	.06302	6.913	502.5	502.5	.5069	.5059	4.955
.5500	.5511	3.463	.9980	15.94	15.97	.9942	.06263	6.925	508.7	508.7	.5068	.5058	4.955
.5510	.5521	3.469	.9981	16.04	16.07	.9942	.06224	6.937	515.0	515.0	.5067	.5058	4.954
.5520	.5531	3.475	.9981	16.14	16.17	.9943	.06186	6.950	521.6	521.6	.5067	.5057	4.954
.5530	.5541	3.481	.9981	16.24	16.27	.9944	.06148	6.962	528.1	528.1	.5066	.5056	4.954
.5540	.5551	3.488	.9981	16.34	16.37	.9944	.06110	6.975	534.8	534.8	.5065	.5056	4.954
.5550	.5560	3.494	.9982	16.44	16.47	.9945	.06073	6.987	541.4	541.4	.5065	.5056	4.953
.5560	.5570	3.500	.9982	16.54	16.57	.9945	.06035	7.000	548.1	548.1	.5064	.5055	4.953
.5570	.5580	3.506	.9982	16.65	16.68	.9946	.05997	7.012	554.9	554.9	.5063	.5054	4.953
.5580	.5590	3.512	.9982	16.75	16.78	.9947	.05960	7.025	562.0	562.0	.5063	.5053	4.953
.5590	.5600	3.519	.9982	16.85	16.88	.9947	.05923	7.037	569.1	569.1	.5062	.5053	4.953
.5600	.5610	3.525	.9983	16.96	16.99	.9947	.05887	7.050	576.1	576.1	.5061	.5053	4.952
.5610	.5620	3.531	.9983	17.06	17.09	.9948	.05850	7.062	583.3	583.3	.5061	.5052	4.952
.5620	.5630	3.537	.9983	17.17	17.20	.9949	.05814	7.074	590.7	590.7	.5060	.5051	4.952
.5630	.5640	3.543	.9983	17.28	17.31	.9949	.05778	7.087	598.0	598.0	.5059	.5051	4.952
.5640	.5649	3.550	.9984	17.38	17.41	.9950	.05743	7.099	605.0	605.0	.5059	.5050	4.951
.5650	.5659	3.556	.9984	17.49	17.52	.9950	.05707	7.112	613.2	613.2	.5058	.5050	4.951
.5660	.5669	3.562	.9984	17.60	17.63	.9951	.05672	7.124	620.8	620.8	.5057	.5049	4.951
.5670	.5679	3.568	.9984	17.71	17.74	.9951	.05637	7.136	628.5	628.5	.5057	.5049	4.951
.5680	.5689	3.575	.9984	17.82	17.85	.9952	.05602	7.149	636.4	636.4	.5056	.5048	4.951
.5690	.5699	3.581	.9985	17.94	17.97	.9952	.05567	7.161	644.3	644.3	.5056	.5048	4.950

Table C–1 – Continued

d/L_o	d/L	2πd/L	TANH 2πd/L	SINH 2πd/L	COSH 2πd/L	H/H'_o	K	4πd/L	SINH 4πd/L	COSH 4πd/L	n	C_G/C_o	M
.5700	.5709	3.587	.9985	18.05	18.08	.9953	.05532	7.174	652.4	652.4	.5055	.5047	4.950
.5710	.5719	3.593	.9985	18.16	18.19	.9953	.05497	7.186	660.5	660.5	.5054	.5047	4.950
.5720	.5729	3.600	.9985	18.28	18.31	.9954	.05463	7.199	668.8	668.8	.5054	.5046	4.950
.5730	.5738	3.606	.9985	18.39	18.42	.9954	.05430	7.211	677.2	677.2	.5053	.5046	4.950
.5740	.5748	3.612	.9985	18.50	18.53	.9955	.05396	7.224	685.6	685.6	.5053	.5045	4.950
.5750	.5758	3.618	.9986	18.62	18.64	.9955	.05363	7.236	694.3	694.3	.5052	.5045	4.949
.5760	.5768	3.624	.9986	18.73	18.76	.9956	.05330	7.249	703.2	703.2	.5052	.5044	4.949
.5770	.5778	3.630	.9986	18.85	18.88	.9956	.05297	7.261	711.9	711.9	.5051	.5044	4.949
.5780	.5788	3.637	.9986	18.97	19.00	.9957	.05264	7.274	720.8	720.8	.5051	.5043	4.949
.5790	.5798	3.643	.9986	19.09	19.12	.9957	.05231	7.286	729.9	729.9	.5050	.5043	4.949
.5800	.5808	3.649	.9987	19.21	19.24	.9957	.05198	7.298	739.0	739.0	.5049	.5043	4.948
.5810	.5818	3.656	.9987	19.33	19.36	.9958	.05166	7.311	748.1	748.1	.5049	.5042	4.948
.5820	.5828	3.662	.9987	19.45	19.48	.9958	.05134	7.323	757.5	757.5	.5048	.5042	4.948
.5830	.5838	3.668	.9987	19.58	19.60	.9959	.05102	7.336	767.0	767.0	.5048	.5041	4.948
.5840	.5848	3.674	.9987	19.70	19.73	.9959	.05070	7.348	776.7	776.7	.5047	.5041	4.948
.5850	.5858	3.680	.9987	19.81	19.84	.9960	.05040	7.361	786.5	786.5	.5047	.5040	4.948
.5860	.5867	3.686	.9987	19.94	19.96	.9960	.05009	7.373	796.4	796.4	.5046	.5040	4.948
.5870	.5877	3.693	.9988	20.06	20.09	.9960	.04978	7.386	806.5	806.5	.5046	.5040	4.947
.5880	.5887	3.699	.9988	20.19	20.21	.9961	.04947	7.398	816.5	816.5	.5045	.5039	4.947
.5890	.5897	3.705	.9988	20.32	20.34	.9961	.04916	7.411	826.7	826.7	.5045	.5039	4.947
.5900	.5907	3.712	.9988	20.45	20.47	.9962	.04885	7.423	837.1	837.1	.5044	.5038	4.947
.5910	.5917	3.718	.9988	20.57	20.60	.9962	.04855	7.436	847.6	847.6	.5044	.5038	4.947
.5920	.5927	3.724	.9988	20.70	20.73	.9963	.04824	7.448	858.2	858.2	.5043	.5037	4.947
.5930	.5937	3.730	.9989	20.83	20.86	.9963	.04794	7.460	868.9	868.9	.5043	.5037	4.946
.5940	.5947	3.737	.9989	20.97	20.99	.9963	.04764	7.473	879.8	879.8	.5043	.5037	4.946
.5950	.5957	3.743	.9989	21.10	21.12	.9964	.04735	7.485	890.8	890.8	.5042	.5036	4.946
.5960	.5967	3.749	.9989	21.23	21.25	.9964	.04706	7.498	901.9	901.9	.5042	.5036	4.946
.5970	.5977	3.755	.9989	21.35	21.37	.9964	.04677	7.510	913.4	913.4	.5041	.5036	4.946
.5980	.5987	3.761	.9989	21.49	21.51	.9965	.04648	7.523	925.0	925.0	.5041	.5035	4.946
.5990	.5996	3.767	.9989	21.62	21.64	.9965	.04619	7.535	936.5	936.5	.5040	.5035	4.946
.6000	.6006	3.774	.9990	21.76	21.78	.9965	.04591	7.548	948.1	948.1	.5040	.5035	4.945
.6100	.6106	3.836	.9991	23.17	23.19	.9969	.04313	7.673	1,074	1,074	.5036	.5031	4.944
.6200	.6205	3.899	.9992	24.66	24.68	.9972	.04052	7.798	1,217	1,217	.5032	.5028	4.943
.6300	.6305	3.961	.9993	26.25	26.27	.9975	.03806	7.923	1,379	1,379	.5029	.5025	4.942
.6400	.6404	4.024	.9994	27.95	27.97	.9977	.03576	8.048	1,527	1,527	.5026	.5023	4.941
.6500	.6504	4.086	.9994	29.75	29.77	.9980	.03359	8.173	1,771	1,771	.5023	.5020	4.940
.6600	.6603	4.149	.9995	31.68	31.69	.9982	.03155	8.298	2,008	2,008	.5021	.5018	4.940
.6700	.6703	4.212	.9996	33.73	33.74	.9983	.02964	8.423	2,275	2,275	.5019	.5017	4.939
.6800	.6803	4.274	.9996	35.90	35.92	.9985	.02784	8.548	2,579	2,579	.5017	.5015	4.939
.6900	.6902	4.337	.9997	38.23	38.24	.9987	.02615	8.674	2,923	2,923	.5015	.5013	4.938
.7000	.7002	4.400	.9997	40.71	40.72	.9988	.02456	8.799	3,314	3,314	.5013	.5012	4.938
.7100	.7102	4.462	.9997	43.34	43.35	.9989	.02307	8.925	3,757	3,757	.5012	.5011	4.937
.7200	.7202	4.525	.9998	46.14	46.15	.9990	.02167	9.050	4,258	4,258	.5011	.5010	4.937
.7300	.7302	4.588	.9998	49.13	49.14	.9991	.02035	9.175	4,828	4,828	.5010	.5009	4.937
.7400	.7401	4.650	.9998	52.31	52.32	.9992	.01911	9.301	5,473	5,473	.5009	.5008	4.937
.7500	.7501	4.713	.9998	55.70	55.71	.9993	.01795	9.426	6,204	6,204	.5008	.5007	4.936
.7600	.7601	4.776	.9999	59.31	59.31	.9994	.01686	9.552	7,034	7,034	.5007	.5006	4.936
.7700	.7701	4.839	.9999	63.15	63.16	.9995	.01583	9.677	7,976	7,976	.5006	.5005	4.936
.7800	.7801	4.902	.9999	67.24	67.25	.9996	.01487	9.803	9,042	9,042	.5005	.5004	4.936
.7900	.7901	4.964	.9999	71.60	71.60	.9996	.01397	9.929	10,250	10,250	.5005	.5004	4.936
.8000	.8001	5.027	.9999	76.24	76.24	.9996	.01312	10.05	11,620	11,620	.5004	.5004	4.936
.8100	.8101	5.090	.9999	81.18	81.19	.9996	.01232	10.18	13,180	13,180	.5004	.5004	4.936
.8200	.8201	5.153	.9999	86.44	86.44	.9997	.01157	10.31	14,940	14,940	.5003	.5003	4.935
.8300	.8301	5.215	.9999	92.04	92.05	.9997	.01086	10.43	17,340	17,340	.5003	.5003	4.935
.8400	.8400	5.278	1.000	98.00	98.01	.9997	.01020	10.56	19,210	19,210	.5003	.5003	4.935
.8500	.8500	5.341	1.000	104.4	104.4	.9998	.009582	10.68	21,780	21,780	.5002	.5002	4.935
.8600	.8600	5.404	1.000	111.1	111.1	.9998	.009000	10.81	24,690	24,690	.5002	.5002	4.935
.8700	.8700	5.467	1.000	118.3	118.3	.9998	.008451	10.93	28,000	28,000	.5002	.5002	4.935
.8800	.8800	5.529	1.000	126.0	126.0	.9998	.007934	11.06	31,750	31,750	.5002	.5002	4.935
.8900	.8900	5.592	1.000	134.2	134.2	.9998	.007454	11.18	36,000	36,000	.5002	.5002	4.935

Table C–1 — Continued

d/L₀	d/L	2πd/L	TANH 2πd/L	SINH 2πd/L	COSH 2πd/L	H/H'₀	K	4πd/L	SINH 4πd/L	COSH 4πd/L	n	C_G/C_0	M
.9000	.9000	5.655	1.000	142.9	142.9	.9999	.007000	11.31	40,810	40,810	.5001	.5001	4.935
.9100	.9100	5.718	1.000	152.1	152.1	.9999	.006574	11.44	46,280	46,280	.5001	.5001	4.935
.9200	.9200	5.781	1.000	162.0	162.0	.9999	.006173	11.56	52,470	52,470	.5001	.5001	4.935
.9300	.9300	5.844	1.000	172.5	172.5	.9999	.005797	11.69	59,500	59,500	.5001	.5001	4.935
.9400	.9400	5.906	1.000	183.7	183.7	.9999	.005445	11.81	67,470	67,470	.5001	.5001	4.935
.9500	.9500	5.969	1.000	195.6	195.6	.9999	.005114	11.94	76,490	76,490	.5001	.5001	4.935
.9600	.9600	6.032	1.000	208.2	208.2	.9999	.004802	12.06	86,740	86,740	.5001	.5001	4.935
.9700	.9700	6.095	1.000	221.7	221.7	.9999	.004510	12.19	98,340	98,340	.5001	.5001	4.935
.9800	.9800	6.158	1.000	236.1	236.1	.9999	.004235	12.32	111,500	111,500	.5001	.5001	4.935
.9900	.9900	6.220	1.000	251.4	251.4	1.000	.003977	12.44	126,500	126,500	.5000	.5000	4.935
1.000	1.000	6.283	1.000	267.7	267.7	1.000	.003735	12.57	143,400	143,400	.5000	.5000	4.935

after Wiegel, R.L., "Oscillatory Waves," U.S. Army, Beach Erosion Board, Bulletin, Special Issue No. 1, July 1948.

Table C–2. Functions of d/L for Even Increments of d/L. (from 0.0001 to 1.000)

d/L	d/L₀	2πd/L	TANH 2πd/L	SINH 2πd/L	COSH 2πd/L	H/H'₀	K	4πd/L	SINH 4πd/L	COSH 4πd/L	n	C_G/C_0	M
0	0	0	0	0	1.0000	∞	1.000	0	0	1.000	1.000	0	∞
0001000	.0000000 6283	.0006283	.0006283	.0006283	1.0000	28.21	1.000	.001257	.001257	1.000	1.000	.0006283	12,500,000
0002000	.000000 2514	.001257	.001257	.001257	1.0000	19.95	1.000	.002513	.002513	1.000	1.000	.001257	3,125,000
0003000	.000000 5655	.001885	.001885	.001885	1.0000	16.29	1.000	.003770	.003770	1.000	1.000	.001885	1,389,000
0004000	.00000 1005	.002513	.002513	.002513	1.0000	14.10	1.000	.005027	.005027	1.000	1.000	.002513	781,300
0005000	.00000 1571	.003142	.003142	.003142	1.0000	12.62	1.000	.006283	.006283	1.000	1.000	.003142	500,000
0006000	.00000 2262	.003770	.003770	.003770	1.0000	11.52	1.000	.007540	.007540	1.000	1.000	.003770	347,200
0007000	.00000 3079	.004398	.004398	.004398	1.0000	10.66	1.000	.008796	.008797	1.000	1.000	.004398	255,100
0008000	.00000 4022	.005027	.005027	.005027	1.0000	9.974	1.000	.01005	.01005	1.000	1.000	.005026	195,300
0009000	.00000 5090	.005655	.005655	.005655	1.0000	9.403	1.000	.01131	.01131	1.000	1.000	.005655	154,300
001000	.00000 6283	.006283	.006283	.006283	1.0000	8.921	1.000	.01257	.01257	1.000	1.000	.006283	125,000
001100	.00000 7603	.006912	.006911	.006912	1.0000	8.506	1.000	.01382	.01382	1.000	1.000	.006911	103,300
001200	.00000 9048	.007540	.007540	.007540	1.0000	8.144	1.000	.01508	.01508	1.000	1.000	.007540	86,810
001300	.00001062	.008168	.008168	.008168	1.0000	7.824	1.000	.01634	.01634	1.000	1.000	.008168	73,970
001400	.00001231	.008796	.008796	.008797	1.0000	7.539	1.000	.01759	.01759	1.000	1.000	.008796	63,780
001500	.00001414	.009425	.009425	.009425	1.0000	7.284	1.000	.01885	.01885	1.000	1.000	.009424	55,560
001600	.00001608	.01005	.01005	.01005	1.0001	7.052	.9999	.02011	.02011	1.000	1.000	.01005	48,830
001700	.00001816	.01068	.01068	.01068	1.0001	6.842	.9999	.02136	.02136	1.000	1.000	.01068	43,260
001800	.00002036	.01131	.01131	.01131	1.0001	6.649	.9999	.02262	.02262	1.000	1.000	.01131	38,580
001900	.00002269	.01194	.01194	.01194	1.0001	6.472	.9999	.02388	.02388	1.000	1.000	.01194	34,630

C-17

Table C—2 — Continued

d/L	d/Lo	$2\pi d/L$	TANH $2\pi d/L$	SINH $2\pi d/L$	COSH $2\pi d/L$	H/H_o'	K	$4\pi d/L$	SINH $4\pi d/L$	COSH $4\pi d/L$	n	C_o/C_o	M
.002000	.00002514	.01257	.01257	.01257	1.0001	6.308	.9999	.02513	.02514	1.000	.9999	.01257	31,250
.002100	.00002772	.01319	.01319	.01320	1.0001	6.156	.9999	.02639	.02639	1.000	.9999	.01319	28,350
.002200	.00003040	.01382	.01382	.01382	1.0001	6.015	.9999	.02765	.02765	1.000	.9999	.01382	25,830
.002300	.00003324	.01445	.01445	.01445	1.0001	5.882	.9999	.02890	.02891	1.000	.9999	.01445	23,630
.002400	.00003619	.01508	.01508	.01508	1.0001	5.759	.9999	.03016	.03016	1.000	.9999	.01508	21,700
.002500	.00003928	.01571	.01571	.01571	1.0001	5.642	.9999	.03142	.03142	1.000	.9999	.01571	20,000
.002600	.00004248	.01634	.01633	.01634	1.0001	5.533	.9999	.03267	.03268	1.001	.9999	.01633	18,490
.002700	.00004579	.01696	.01696	.01697	1.0001	5.429	.9999	.03393	.03394	1.001	.9999	.01696	17,150
.002800	.00004925	.01759	.01759	.01759	1.0002	5.332	.9998	.03519	.03519	1.001	.9999	.01759	15,950
.002900	.00005284	.01822	.01822	.01822	1.0002	5.239	.9998	.03644	.03645	1.001	.9999	.01822	14,870
.003000	.00005652	.01885	.01885	.01885	1.0002	5.151	.9998	.03770	.03771	1.001	.9999	.01885	13,890
.003100	.00006039	.01948	.01948	.01948	1.0002	5.067	.9998	.03896	.03897	1.001	.9999	.01947	13,010
.003200	.00006435	.02011	.02010	.02011	1.0002	4.987	.9998	.04021	.04022	1.001	.9999	.02010	12,210
.003300	.00006841	.02073	.02073	.02073	1.0002	4.911	.9998	.04147	.04148	1.001	.9999	.02073	11,480
.003400	.00007262	.02136	.02136	.02136	1.0002	4.838	.9998	.04273	.04274	1.001	.9998	.02136	10,820
.003500	.00007697	.02199	.02199	.02199	1.0002	4.769	.9998	.04398	.04399	1.001	.9998	.02199	10,210
.003600	.00008140	.02262	.02262	.02262	1.0003	4.702	.9997	.04524	.04525	1.001	.9998	.02261	9,648
.003700	.00008599	.02325	.02324	.02325	1.0003	4.638	.9997	.04650	.04652	1.001	.9998	.02324	9,134
.003800	.00009071	.02388	.02387	.02388	1.0003	4.577	.9997	.04775	.04777	1.001	.9998	.02387	8,660
.033900	.00009551	.02450	.02450	.02451	1.0003	4.518	.9997	.04901	.04903	1.001	.9998	.02449	8,221
.004000	.0001005	.02513	.02513	.02513	1.0003	4.462	.9997	.05027	.05029	1.001	.9998	.02511	7,815
.004100	.0001056	.02576	.02576	.02576	1.0003	4.407	.9997	.05152	.05154	1.001	.9998	.02574	7,439
.004200	.0001108	.02639	.02638	.02639	1.0003	4.354	.9997	.05278	.05280	1.001	.9998	.02637	7,090
.004300	.0001161	.02702	.02701	.02702	1.0004	4.303	.9996	.05404	.05406	1.001	.9998	.02700	6,764
.004400	.0001216	.02765	.02764	.02765	1.0004	4.254	.9996	.05529	.05531	1.002	.9997	.02763	6,460
.004500	.0001272	.02827	.02827	.02828	1.0004	4.207	.9996	.05655	.05658	1.002	.9997	.02825	6,176
.004600	.0001329	.02890	.02889	.02890	1.0004	4.161	.9996	.05781	.05784	1.002	.9997	.02888	5,911
.004700	.0001387	.02953	.02952	.02953	1.0004	4.116	.9996	.05906	.05909	1.002	.9997	.02951	5,662
.004800	.0001447	.03016	.03015	.03016	1.0005	4.073	.9995	.06032	.06035	1.002	.9997	.03014	5,429
.004900	.0001508	.03079	.03078	.03079	1.0005	4.032	.9995	.06158	.06161	1.002	.9997	.03076	5,209
.005000	.0001570	.03142	.03141	.03143	1.0005	3.991	.9995	.06283	.06287	1.002	.9997	.03139	5,003
.005100	.0001634	.03204	.03203	.03205	1.0005	3.951	.9995	.06409	.06413	1.002	.9997	.03202	4,809
.005200	.0001698	.03267	.03266	.03268	1.0005	3.913	.9995	.06535	.06539	1.002	.9996	.03265	4,626
.005300	.0001764	.03330	.03329	.03331	1.0005	3.876	.9995	.06660	.06665	1.002	.9996	.03328	4,453
.005400	.0001832	.03393	.03392	.03394	1.0006	3.840	.9994	.06786	.06791	1.002	.9996	.03391	4,290
.005500	.0001900	.03456	.03455	.03457	1.0006	3.805	.9994	.06911	.06916	1.002	.9996	.03454	4,135
.005600	.0001970	.03519	.03517	.03520	1.0006	3.771	.9994	.07037	.07042	1.002	.9996	.03517	3,989
.005700	.0002041	.03581	.03580	.03582	1.0006	3.738	.9994	.07163	.07169	1.003	.9996	.03579	3,851
.005800	.0002112	.03644	.03642	.03645	1.0007	3.706	.9993	.07288	.07294	1.003	.9996	.03641	3,719
.005900	.0002186	.03707	.03705	.03708	1.0007	3.675	.9993	.07414	.07420	1.003	.9995	.03703	3,594
.006000	.0002261	.03770	.03768	.03771	1.0007	3.644	.9993	.07540	.07547	1.003	.9995	.03766	3,475
.006100	.0002337	.03833	.03831	.03834	1.0007	3.614	.9993	.07665	.07672	1.003	.9995	.03829	3,363
.006200	.0002414	.03896	.03894	.03897	1.0008	3.584	.9992	.07791	.07798	1.003	.9995	.03892	3,255
.006300	.0002492	.03958	.03956	.03959	1.0008	3.556	.9992	.07917	.07925	1.003	.9995	.03954	3,153
.006400	.0002570	.04021	.04019	.04022	1.0008	3.528	.9992	.08042	.08050	1.003	.9995	.04017	3,055
.006500	.0002653	.04084	.04082	.04085	1.0008	3.501	.9992	.08168	.08177	1.003	.9994	.04080	2,962
.006600	.0002735	.04147	.04144	.04148	1.0009	3.475	.9991	.08294	.08303	1.003	.9994	.04142	2,873
.006700	.0002819	.04210	.04207	.04211	1.0009	3.449	.9991	.08419	.08428	1.004	.9994	.04204	2,788
.006800	.0002904	.04273	.04270	.04274	1.0009	3.423	.9991	.08545	.08555	1.004	.9994	.04267	2,707
.006900	.0002990	.04335	.04333	.04336	1.0009	3.398	.9991	.08671	.08681	1.004	.9994	.04330	2,629
.007000	.0003077	.04398	.04395	.04399	1.0010	3.374	.9990	.08796	.08807	1.004	.9994	.04392	2,554
.007100	.0003165	.04461	.04458	.04462	1.0010	3.350	.9990	.08922	.08933	1.004	.9993	.04455	2,483
.007200	.0003254	.04524	.04521	.04525	1.0010	3.327	.9989	.09048	.09060	1.004	.9993	.04518	2,415
.007300	.0003346	.04587	.04584	.04589	1.0011	3.304	.9989	.09173	.09185	1.004	.9993	.04581	2,349
.007400	.0003439	.04650	.04646	.04652	1.0011	3.281	.9989	.09299	.09312	1.004	.9993	.04644	2,286
.007500	.0003532	.04712	.04709	.04714	1.0011	3.260	.9989	.09425	.09438	1.004	.9993	.04706	2,226
.007600	.0003627	.04775	.04772	.04777	1.0011	3.238	.9989	.09550	.09565	1.005	.9992	.04768	2,167
.007700	.0003722	.04838	.04834	.04840	1.0012	3.217	.9988	.09676	.09681	1.005	.9992	.04830	2,112
.007800	.0003820	.04901	.04897	.04903	1.0012	3.197	.9988	.09802	.09817	1.005	.9992	.04893	2,058
.007900	.0003918	.04964	.04960	.04966	1.0012	3.176	.9988	.09927	.09943	1.005	.9992	.04956	2,006

Table C—2 — Continued

d/L	d/Lo	2πd/L	TANH 2πd/L	SINH 2πd/L	COSH 2πd/L	H/H'o	K	4πd/L	SINH 4πd/L	COSH 4πd/L	n	Co/Co	M
.008000	.0004018	.05027	.05022	.05029	1.0013	3.157	.9987	.1005	.1007	1.005	.9992	.05018	1,956
.008100	.0004118	.05089	.05085	.05091	1.0013	3.137	.9987	.1018	.1020	1.005	.9991	.05080	1,909
.008200	.0004221	.05152	.05147	.05154	1.0013	3.118	.9987	.1030	.1032	1.005	.9991	.05142	1,862
.008300	.0004324	.05215	.05210	.05217	1.0014	3.099	.9986	.1043	.1045	1.005	.9991	.05205	1,818
.008400	.0004429	.05278	.05273	.05280	1.0014	3.081	.9986	.1056	.1058	1.006	.9991	.05268	1,775
.008500	.0004536	.05341	.05336	.05343	1.0014	3.062	.9986	.1068	.1070	1.006	.9991	.05331	1,733
.008600	.0004644	.05404	.05398	.05406	1.0015	3.044	.9985	.1081	.1083	1.006	.9990	.05394	1,693
.008700	.0004751	.05466	.05461	.05469	1.0015	3.027	.9985	.1093	.1095	1.006	.9990	.05456	1,655
.008800	.0004860	.05529	.05524	.05533	1.0015	3.010	.9985	.1106	.1108	1.006	.9990	.05518	1,617
.008900	.0004972	.05592	.05586	.05595	1.0016	2.993	.9984	.1118	.1121	1.006	.9990	.05580	1,581
.009000	.0005084	.05655	.05649	.05658	1.0016	2.977	.9984	.1131	.1133	1.006	.9989	.05643	1,546
.009100	.0005198	.05718	.05712	.05721	1.0016	2.960	.9984	.1144	.1146	1.006	.9989	.05706	1,513
.009200	.0005312	.05781	.05774	.05784	1.0017	2.944	.9983	.1156	.1158	1.007	.9989	.05768	1,480
.009300	.0005427	.05843	.05836	.05846	1.0017	2.929	.9983	.1169	.1171	1.007	.9989	.05830	1,449
.009400	.0005545	.05906	.05899	.05909	1.0017	2.913	.9983	.1181	.1184	1.007	.9988	.05892	1,418
.009500	.0005664	.05969	.05962	.05973	1.0018	2.898	.9982	.1194	.1196	1.007	.9988	.05955	1,388
.009600	.0005784	.06032	.06025	.06036	1.0018	2.882	.9982	.1206	.1209	1.007	.9988	.06018	1,360
.009700	.0005905	.06095	.06087	.06099	1.0019	2.867	.9981	.1219	.1222	1.007	.9988	.06080	1,332
.009800	.0006027	.06158	.06150	.06162	1.0019	2.853	.9981	.1232	.1235	1.008	.9987	.06142	1,305
.009900	.0006150	.06220	.06212	.06224	1.0019	2.839	.9981	.1244	.1247	1.008	.9987	.06204	1,279
.01000	.0006275	.06283	.06275	.06287	1.0020	2.825	.9980	.1257	.1260	1.0079	.9987	.06267	1,253
.01100	.0007591	.06912	.06901	.06917	1.0024	2.694	.9976	.1382	.1387	1.0096	.9984	.06890	1,036
.01200	.0009031	.07540	.07526	.07547	1.0028	2.580	.9972	.1508	.1513	1.0114	.9981	.07511	871.0
.01300	.001060	.08168	.08150	.08177	1.0033	2.480	.9967	.1634	.1641	1.0134	.9978	.08131	742.9
.01400	.001228	.08795	.08774	.08808	1.0039	2.389	.9961	.1759	.1768	1.0155	.9974	.08751	641.1
.01500	.001410	.09425	.09397	.09439	1.0044	2.310	.9956	.1885	.1896	1.0178	.9970	.09369	558.9
.01600	.001603	.1005	.1002	.1007	1.0051	2.238	.9949	.2011	.2024	1.0203	.9966	.09986	491.6
.01700	.001809	.1068	.1064	.1070	1.0057	2.172	.9943	.2136	.2153	1.0229	.9962	.1060	435.8
.01800	.002027	.1131	.1126	.1133	1.0064	2.112	.9936	.2262	.2281	1.0257	.9958	.1121	389.1
.01900	.002258	.1194	.1188	.1197	1.0071	2.056	.9929	.2388	.2410	1.0286	.9953	.1183	349.5
.02000	.002500	.1257	.1250	.1260	1.008	2.005	.9922	.2513	.2540	1.032	.9947	.1244	315.8
.02100	.002755	.1320	.1312	.1323	1.009	1.958	.9914	.2639	.2669	1.035	.9942	.1305	286.8
.02200	.003022	.1382	.1374	.1387	1.010	1.915	.9905	.2765	.2800	1.038	.9937	.1365	261.5
.02300	.003301	.1445	.1435	.1450	1.011	1.873	.9896	.2890	.2931	1.042	.9931	.1425	239.6
.02400	.003592	.1508	.1497	.1514	1.011	1.834	.9887	.3016	.3062	1.046	.9925	.1485	220.3
.02500	.003895	.1571	.1558	.1577	1.012	1.799	.9878	.3142	.3194	1.050	.9919	.1545	203.3
.02600	.004210	.1634	.1619	.1641	1.013	1.765	.9868	.3267	.3326	1.054	.9912	.1605	188.2
.02700	.004537	.1697	.1680	.1705	1.014	1.733	.9858	.3458	.3458	1.058	.9905	.1665	174.8
.02800	.004876	.1759	.1741	.1768	1.016	1.703	.9847	.3519	.3592	1.063	.9898	.1724	162.7
.02900	.005226	.1822	.1802	.1832	1.017	1.675	.9836	.3644	.3725	1.067	.9891	.1783	151.9
.03000	.005589	.1885	.1863	.1896	1.018	1.648	.9825	.3770	.3860	1.072	.9884	.1841	142.2
.03100	.005963	.1948	.1924	.1960	1.019	1.622	.9813	.3896	.3995	1.077	.9876	.1900	133.4
.03200	.006347	.2011	.1984	.2024	1.020	1.598	.9801	.4021	.4131	1.082	.9868	.1958	125.4
.03300	.006746	.2073	.2044	.2088	1.022	1.575	.9789	.4147	.4267	1.087	.9860	.2016	118.1
.03400	.007155	.2136	.2104	.2153	1.023	1.553	.9776	.4273	.4404	1.093	.9851	.2073	111.4
.03500	.007575	.2199	.2164	.2217	1.024	1.532	.9763	.4398	.4541	1.098	.9843	.2130	105.3
.03600	.008007	.2262	.2224	.2281	1.026	1.512	.9749	.4524	.4680	1.104	.9834	.2187	99.75
.03700	.008450	.2325	.2284	.2346	1.027	1.493	.9736	.4650	.4819	1.110	.9824	.2244	94.61
.03800	.008905	.2388	.2343	.2410	1.029	1.475	.9722	.4775	.4959	1.116	.9815	.2300	89.88
.03900	.009370	.2450	.2403	.2527	1.030	1.457	.9708	.4901	.5099	1.123	.9805	.2356	85.50
.04000	.009847	.2513	.2462	.2540	1.032	1.440	.9693	.5027	.5241	1.129	.9795	.2411	81.43
.04100	.01033	.2576	.2521	.2605	1.033	1.424	.9677	.5152	.5383	1.136	.9785	.2467	77.67
.04200	.01083	.2639	.2579	.2670	1.035	1.408	.9662	.5278	.5526	1.143	.9775	.2521	74.17
.04300	.01134	.2702	.2638	.2735	1.037	1.393	.9646	.5404	.5670	1.150	.9765	.2576	70.91
.04400	.01186	.2765	.2696	.2800	1.039	1.379	.9630	.5529	.5815	1.157	.9754	.2630	67.88
.04500	.01239	.2827	.2754	.2865	1.040	1.365	.9613	.5655	.5961	1.164	.9743	.2684	65.05
.04600	.01294	.2890	.2812	.2931	1.042	1.352	.9596	.5781	.6108	1.172	.9732	.2737	62.39
.04700	.01349	.2953	.2870	.2996	1.044	1.339	.9579	.5906	.6256	1.180	.9721	.2790	59.91
.04800	.01405	.3016	.2928	.3062	1.046	1.326	.9562	.6032	.6404	1.188	.9709	.2843	57.57
.04900	.01463	.3079	.2985	.3128	1.048	1.314	.9544	.6158	.6554	1.196	.9697	.2895	55.38

Table C–2 — Continued

d/L	d/Lo	2πd/L	TANH 2πd/L	SINH 2πd/L	COSH 2πd/L	H/H'o	K	4πd/L	SINH 4πd/L	COSH 4πd/L	n	CG/Co	M
.05000	.01521	.3142	.3042	.3194	1.050	1.303	.9526	.6283	.6705	1.204	.9685	.2947	53.32
.05100	.01580	.3204	.3099	.3260	1.052	1.291	.9508	.6409	.6857	1.213	.9673	.2998	51.38
.05200	.01641	.3267	.3156	.3326	1.054	1.281	.9489	.6535	.7010	1.221	.9661	.3049	49.55
.05300	.01702	.3330	.3212	.3392	1.056	1.270	.9470	.6660	.7164	1.230	.9649	.3099	47.82
.05400	.01765	.3393	.3269	.3458	1.058	1.260	.9451	.6786	.7319	1.239	.9636	.3149	46.19
.05500	.01829	.3456	.3325	.3525	1.060	1.250	.9431	.6912	.7475	1.249	.9623	.3199	44.65
.05600	.01893	.3519	.3380	.3592	1.063	1.241	.9411	.7037	.7633	1.258	.9610	.3248	43.19
.05700	.01958	.3581	.3436	.3658	1.065	1.231	.9391	.7163	.7791	1.268	.9597	.3297	41.80
.05800	.02025	.3644	.3491	.3726	1.067	1.222	.9371	.7289	.7951	1.278	.9583	.3346	40.49
.05900	.02092	.3707	.3546	.3793	1.070	1.214	.9350	.7414	.8112	1.288	.9570	.3394	39.24
.06000	.02161	.3770	.3601	.3860	1.072	1.205	.9329	.7540	.8275	1.298	.9556	.3441	38.06
.06100	.02230	.3833	.3656	.3927	1.074	1.197	.9308	.7666	.8439	1.308	.9542	.3488	36.93
.06200	.02300	.3896	.3710	.3995	1.077	1.189	.9286	.7791	.8604	1.319	.9528	.3534	35.86
.06300	.02371	.3958	.3764	.4062	1.079	1.182	.9265	.7917	.8770	1.330	.9514	.3581	34.83
.06400	.02444	.4021	.3818	.4130	1.082	1.174	.9243	.8043	.8938	1.341	.9499	.3626	33.86
.06500	.02516	.4084	.3871	.4199	1.085	1.167	.9220	.8168	.9107	1.353	.9484	.3672	32.93
.06600	.02590	.4147	.3925	.4267	1.087	1.160	.9198	.8294	.9278	1.364	.9470	.3716	32.04
.06700	.02665	.4210	.3978	.4335	1.090	1.153	.9175	.8419	.9450	1.376	.9455	.3761	31.19
.06800	.02739	.4273	.4030	.4404	1.093	1.147	.9152	.8545	.9624	1.388	.9440	.3804	30.38
.06900	.02817	.4335	.4083	.4473	1.095	1.140	.9128	.8671	.9799	1.400	.9424	.3848	29.61
.07000	.02895	.4398	.4135	.4541	1.098	1.134	.9105	.8796	.9976	1.412	.9409	.3891	28.86
.07100	.02973	.4461	.4187	.4611	1.101	1.128	.9081	.8922	1.015	1.425	.9393	.3933	28.15
.07200	.03052	.4524	.4239	.4680	1.104	1.122	.9057	.9048	1.033	1.438	.9378	.3975	27.47
.07300	.03132	.4587	.4290	.4749	1.107	1.116	.9033	.9173	1.052	1.451	.9362	.4016	26.81
.07400	.03213	.4650	.4341	.4819	1.110	1.110	.9008	.9299	1.070	1.464	.9346	.4057	26.18
.07500	.03294	.4712	.4392	.4889	1.113	1.105	.8984	.9425	1.088	1.478	.9330	.4098	25.58
.07600	.03377	.4775	.4443	.4958	1.116	1.099	.8959	.9551	1.107	1.492	.9314	.4138	25.00
.07700	.03460	.4838	.4493	.5029	1.119	1.094	.8934	.9676	1.126	1.506	.9298	.4177	24.45
.07800	.03543	.4901	.4542	.5100	1.123	1.089	.8909	.9802	1.145	1.520	.9281	.4216	23.92
.07900	.03628	.4964	.4593	.5170	1.126	1.084	.8883	.9927	1.164	1.534	.9264	.4255	23.40
.08000	.03714	.5027	.4642	.5241	1.129	1.079	.8857	1.005	1.183	1.549	.9248	.4293	22.90
.08100	.03799	.5089	.4691	.5312	1.132	1.075	.8831	1.018	1.203	1.564	.9231	.4330	22.42
.08200	.03887	.5152	.4740	.5383	1.136	1.070	.8805	1.030	1.223	1.580	.9214	.4367	21.96
.08300	.03975	.5215	.4789	.5455	1.139	1.066	.8779	1.043	1.243	1.595	.9197	.4404	21.52
.08400	.04063	.5278	.4837	.5526	1.143	1.061	.8752	1.056	1.263	1.611	.9179	.4440	21.09
.08500	.04152	.5341	.4885	.5598	1.146	1.057	.8726	1.068	1.283	1.627	.9162	.4476	20.68
.08600	.04242	.5404	.4933	.5670	1.150	1.053	.8699	1.081	1.304	1.643	.9145	.4511	20.28
.08700	.04333	.5466	.4980	.5743	1.153	1.049	.8672	1.093	1.324	1.660	.9127	.4545	19.90
.08800	.04424	.5529	.5027	.5815	1.157	1.045	.8645	1.106	1.346	1.676	.9109	.4579	19.53
.08900	.04516	.5592	.5074	.5888	1.160	1.041	.8617	1.118	1.367	1.693	.9092	.4613	19.17
.09000	.04608	.5655	.5120	.5961	1.164	1.037	.8590	1.131	1.388	1.711	.9074	.4646	18.82
.09100	.04702	.5718	.5167	.6034	1.168	1.034	.8562	1.144	1.410	1.728	.9056	.4679	18.49
.09200	.04796	.5781	.5213	.6108	1.172	1.030	.8534	1.156	1.431	1.746	.9038	.4711	18.16
.09300	.04890	.5843	.5258	.6182	1.176	1.027	.8506	1.169	1.453	1.764	.9020	.4743	17.85
.09400	.04985	.5906	.5303	.6256	1.180	1.023	.8478	1.181	1.476	1.783	.9002	.4774	17.55
.09500	.05081	.5969	.5348	.6330	1.184	1.020	.8450	1.194	1.498	1.801	.8984	.4805	17.26
.09600	.05177	.6032	.5393	.6404	1.188	1.017	.8421	1.206	1.521	1.820	.8966	.4835	16.97
.09700	.05275	.6095	.5438	.6479	1.192	1.014	.8392	1.219	1.544	1.840	.8947	.4865	16.69
.09800	.05372	.6158	.5482	.6554	1.196	1.011	.8364	1.232	1.567	1.859	.8929	.4894	16.42
.09900	.05470	.6220	.5526	.6629	1.200	1.008	.8335	1.244	1.591	1.879	.8910	.4923	16.16
.1000	.05569	.6283	.5569	.6705	1.204	1.005	.8306	1.257	1.615	1.899	.8892	.4952	15.91
.1010	.05668	.6346	.5612	.6781	1.208	1.002	.8277	1.269	1.638	1.920	.8873	.4980	15.67
.1020	.05768	.6409	.5655	.6857	1.213	.9993	.8247	1.282	1.663	1.940	.8854	.5007	15.43
.1030	.05869	.6472	.5698	.6933	1.217	.9966	.8218	1.294	1.687	1.961	.8836	.5034	15.20
.1040	.05970	.6535	.5740	.7010	1.221	.9940	.8189	1.307	1.712	1.983	.8817	.5061	14.98
.1050	.06071	.6597	.5782	.7087	1.226	.9914	.8159	1.319	1.737	2.004	.8798	.5087	14.76
.1060	.06173	.6660	.5824	.7164	1.230	.9891	.8129	1.332	1.762	2.026	.8779	.5113	14.55
.1070	.06276	.6723	.5865	.7241	1.235	.9865	.8100	1.345	1.788	2.049	.8760	.5138	14.35
.1080	.06378	.6786	.5906	.7319	1.239	.9841	.8070	1.357	1.814	2.071	.8741	.5163	14.15
.1090	.06482	.6849	.5947	.7397	1.244	.9818	.8040	1.370	1.840	2.094	.8722	.5187	13.95

Table C–2 – Continued

d/L	d/Lo	2πd/L	TANH 2πd/L	SINH 2πd/L	COSH 2πd/L	H/H'o	K	4πd/L	SINH 4πd/L	COSH 4πd/L	n	Co/Co	M
.1100	.06586	.6912	.5987	.7475	1.249	.9197	.8010	1.382	1.867	2.118	.8703	.5211	13.77
.1110	.06690	.6974	.6027	.7554	1.253	.9775	.7980	1.395	1.893	2.141	.8684	.5234	13.58
.1120	.06795	.7037	.6067	.7633	1.258	.9753	.7949	1.407	1.920	2.165	.8665	.5257	13.41
.1130	.06901	.7100	.6107	.7712	1.263	.9731	.7919	1.420	1.948	2.189	.8645	.5279	13.23
.1140	.07006	.7163	.6146	.7791	1.268	.9711	.7888	1.433	1.975	2.214	.8626	.5301	13.06
.1150	.07113	.7226	.6185	.7871	1.273	.9691	.7858	1.445	2.003	2.239	.8607	.5323	12.90
.1160	.07220	.7289	.6224	.7951	1.278	.9672	.7827	1.458	2.032	2.264	.8587	.5344	12.74
.1170	.07327	.7351	.6262	.8032	1.283	.9654	.7797	1.470	2.060	2.290	.8568	.5365	12.59
.1180	.07434	.7414	.6300	.8112	1.288	.9635	.7766	1.483	2.089	2.316	.8549	.5386	12.43
.1190	.07542	.7477	.6338	.8193	1.293	.9617	.7735	1.495	2.118	2.343	.8529	.5406	12.29
.1200	.07650	.7540	.6375	.8275	1.298	.9600	.7704	1.508	2.148	2.369	.8510	.5425	12.14
.1210	.07759	.7603	.6412	.8357	1.303	.9583	.7673	1.521	2.178	2.397	.8491	.5444	12.00
.1220	.07868	.7666	.6449	.8439	1.309	.9567	.7642	1.533	2.208	2.424	.8471	.5463	11.87
.1230	.07978	.7728	.6486	.8521	1.314	.9551	.7612	1.546	2.239	2.452	.8452	.5482	11.73
.1240	.08085	.7791	.6520	.8604	1.319	.9535	.7581	1.558	2.270	2.480	.8432	.5500	11.61
.1250	.08198	.7854	.6558	.8687	1.325	.9520	.7549	1.571	2.301	2.509	.8413	.5517	11.48
.1260	.08308	.7917	.6594	.8770	1.330	.9505	.7518	1.583	2.333	2.538	.8393	.5534	11.35
.1270	.08419	.7980	.6629	.8854	1.336	.9490	.7487	1.596	2.365	2.568	.8374	.5551	11.23
.1280	.08530	.8043	.6664	.8938	1.341	.9476	.7456	1.609	2.398	2.598	.8354	.5568	11.11
.1290	.08642	.8105	.6699	.9022	1.347	.9463	.7424	1.621	2.430	2.628	.8335	.5584	11.00
.1300	.08753	.8168	.6733	.9107	1.353	.9450	.7393	1.634	2.464	2.659	.8316	.5599	10.89
.1310	.08866	.8231	.6768	.9192	1.358	.9437	.7362	1.646	2.497	2.690	.8296	.5614	10.78
.1320	.08978	.8294	.6801	.9278	1.364	.9424	.7331	1.659	2.531	2.722	.8277	.5629	10.67
.1330	.09091	.8357	.6835	.9364	1.370	.9412	.7299	1.671	2.566	2.754	.8257	.5644	10.56
.1340	.09204	.8420	.6868	.9450	1.376	.9401	.7268	1.684	2.600	2.786	.8238	.5658	10.46
.1350	.09317	.8482	.6902	.9537	1.382	.9389	.7237	1.696	2.636	2.819	.8218	.5672	10.36
.1360	.09431	.8545	.6934	.9624	1.388	.9378	.7205	1.709	2.671	2.852	.8199	.5685	10.26
.1370	.09544	.8608	.6967	.9711	1.394	.9367	.7174	1.722	2.707	2.886	.8179	.5698	10.17
.1380	.09659	.8671	.6999	.9799	1.400	.9357	.7142	1.734	2.744	2.920	.8160	.5711	10.07
.1390	.09773	.8734	.7031	.9887	1.406	.9347	.7111	1.747	2.781	2.955	.8141	.5724	9.983
.1400	.09888	.8797	.7063	.9976	1.412	.9337	.7080	1.759	2.818	2.990	.8121	.5736	9.894
.1410	.1000	.8859	.7094	1.006	1.419	.9327	.7048	1.772	2.856	3.026	.8102	.5748	9.806
.1420	.1012	.8922	.7125	1.015	1.425	.9318	.7017	1.784	2.894	3.062	.8083	.5759	9.721
.1430	.1023	.8985	.7156	1.024	1.432	.9309	.6985	1.797	2.933	3.099	.8064	.5770	9.638
.1440	.1035	.9048	.7186	1.033	1.438	.9300	.6954	1.810	2.972	3.136	.8044	.5781	9.556
.1450	.1046	.9111	.7216	1.042	1.445	.9292	.6923	1.822	3.012	3.173	.8025	.5791	9.476
.1460	.1058	.9174	.7247	1.052	1.451	.9284	.6891	1.835	3.052	3.211	.8006	.5801	9.398
.1470	.1070	.9236	.7276	1.061	1.458	.9276	.6860	1.847	3.092	3.250	.7987	.5811	9.321
.1480	.1081	.9299	.7306	1.070	1.464	.9268	.6829	1.860	3.133	3.289	.7968	.5821	9.246
.1490	.1093	.9362	.7335	1.079	1.471	.9261	.6797	1.872	3.175	3.329	.7949	.5830	9.173
.1500	.1105	.9425	.7364	1.088	1.478	.9254	.6766	1.885	3.217	3.369	.7930	.5839	9.101
.1510	.1116	.9488	.7392	1.098	1.485	.9247	.6734	1.898	3.260	3.410	.7911	.5848	9.031
.1520	.1128	.9551	.7421	1.107	1.492	.9240	.6703	1.910	3.303	3.451	.7892	.5856	8.962
.1530	.1140	.9613	.7449	1.116	1.499	.9234	.6672	1.923	3.346	3.493	.7873	.5864	8.894
.1540	.1151	.9676	.7477	1.126	1.506	.9228	.6641	1.935	3.391	3.535	.7854	.5872	8.828
.1550	.1163	.9739	.7504	1.135	1.513	.9222	.6610	1.948	3.435	3.578	.7835	.5880	8.763
.1560	.1175	.9802	.7531	1.145	1.520	.9216	.6579	1.960	3.481	3.621	.7816	.5887	8.700
.1570	.1187	.9865	.7558	1.154	1.527	.9211	.6547	1.973	3.526	3.665	.7797	.5893	8.638
.1580	.1199	.9928	.7585	1.164	1.535	.9205	.6516	1.985	3.573	3.710	.7779	.5900	8.577
.1590	.1210	.9990	.7612	1.174	1.542	.9200	.6485	1.998	3.620	3.755	.7760	.5907	8.517
.1600	.1222	1.005	.7638	1.183	1.549	.9196	.6454	2.011	3.667	3.801	.7741	.5913	8.459
.1610	.1234	1.012	.7664	1.193	1.557	.9191	.6423	2.023	3.715	3.847	.7723	.5919	8.401
.1620	.1246	1.018	.7690	1.203	1.564	.9186	.6392	2.036	3.764	3.894	.7704	.5925	8.345
.1630	.1258	1.024	.7716	1.213	1.572	.9182	.6361	2.048	3.813	3.942	.7686	.5930	8.290
.1640	.1270	1.030	.7741	1.223	1.580	.9179	.6331	2.061	3.863	3.990	.7667	.5935	8.236
.1650	.1281	1.037	.7766	1.233	1.587	.9175	.6300	2.073	3.913	4.039	.7649	.5940	8.183
.1660	.1293	1.043	.7791	1.243	1.595	.9171	.6269	2.086	3.964	4.088	.7631	.5945	8.131
.1670	.1305	1.049	.7815	1.253	1.603	.9167	.6239	2.099	4.016	4.138	.7613	.5950	8.079
.1680	.1317	1.056	.7840	1.263	1.611	.9164	.6208	2.111	4.068	4.189	.7595	.5954	8.029
.1690	.1329	1.062	.7864	1.273	1.619	.9161	.6177	2.124	4.121	4.241	.7576	.5958	7.980

Table C–2 – Continued

d/L	d/Lo	2πd/L	TANH 2πd/L	SINH 2πd/L	COSH 2πd/L	H/H'o	K	4πd/L	SINH 4πd/L	COSH 4πd/L	n	Co/Co	M
.1700	.1341	1.068	.7887	1.283	1.627	.9158	.6147	2.136	4.175	4.293	.7558	.5962	7.932
.1710	.1353	1.074	.7911	1.293	1.635	.9155	.6117	2.149	4.229	4.346	.7540	.5965	7.885
.1720	.1365	1.081	.7935	1.304	1.643	.9153	.6086	2.161	4.284	4.399	.7523	.5969	7.838
.1730	.1377	1.087	.7958	1.314	1.651	.9150	.6056	2.174	4.340	4.454	.7505	.5972	7.793
.1740	.1389	1.093	.7981	1.325	1.660	.9148	.6026	2.187	4.396	4.508	.7487	.5975	7.748
.1750	.1401	1.100	.8004	1.335	1.668	.9146	.5995	2.199	4.453	4.564	.7469	.5978	7.704
.1760	.1413	1.106	.8026	1.345	1.676	.9144	.5965	2.212	4.511	4.620	.7451	.5980	7.661
.1770	.1425	1.112	.8048	1.356	1.685	.9142	.5935	2.224	4.569	4.677	.7434	.5983	7.619
.1780	.1437	1.118	.8070	1.367	1.693	.9140	.5905	2.237	4.628	4.735	.7416	.5985	7.577
.1790	.1449	1.125	.8092	1.377	1.702	.9138	.5875	2.249	4.688	4.793	.7399	.5987	7.536
.1800	.1460	1.131	.8114	1.388	1.711	.9137	.5845	2.262	4.749	4.853	.7382	.5989	7.496
.1810	.1472	1.137	.8135	1.399	1.720	.9136	.5816	2.275	4.810	4.918	.7364	.5991	7.457
.1820	.1484	1.144	.8156	1.410	1.728	.9135	.5786	2.287	4.872	4.974	.7347	.5992	7.419
.1830	.1496	1.150	.8177	1.420	1.737	.9134	.5757	2.300	4.935	5.035	.7330	.5993	7.381
.1840	.1508	1.156	.8198	1.431	1.746	.9133	.5727	2.312	4.999	5.098	.7313	.5995	7.343
.1850	.1520	1.162	.8218	1.442	1.755	.9132	.5697	2.325	5.063	5.161	.7296	.5996	7.307
.1860	.1532	1.169	.8239	1.454	1.764	.9131	.5668	2.337	5.129	5.225	.7279	.5997	7.271
.1870	.1544	1.175	.8259	1.465	1.773	.9131	.5639	2.350	5.195	5.290	.7262	.5997	7.235
.1880	.1556	1.181	.8278	1.476	1.783	.9131	.5610	2.362	5.262	5.356	.7245	.5998	7.201
.1890	.1568	1.188	.8298	1.487	1.792	.9130	.5581	2.375	5.329	5.422	.7228	.5998	7.167
.1900	.1580	1.194	.8318	1.498	1.801	.9130	.5551	2.388	5.398	5.490	.7212	.5998	7.133
.1910	.1592	1.200	.8337	1.510	1.811	.9130	.5522	2.400	5.467	5.558	.7195	.5998	7.100
.1920	.1604	1.206	.8356	1.521	1.820	.9130	.5493	2.413	5.538	5.625	.7179	.5998	7.068
.1930	.1616	1.213	.8375	1.533	1.830	.9130	.5465	2.425	5.609	5.697	.7162	.5998	7.036
.1940	.1628	1.219	.8393	1.544	1.840	.9131	.5436	2.438	5.681	5.768	.7146	.5998	7.005
.1950	.1640	1.225	.8412	1.556	1.849	.9131	.5408	2.450	5.754	5.840	.7129	.5997	6.974
.1960	.1652	1.232	.8430	1.567	1.859	.9131	.5379	2.463	5.827	5.913	.7113	.5997	6.944
.1970	.1664	1.238	.8448	1.579	1.869	.9132	.5350	2.476	5.902	5.988	.7097	.5996	6.914
.1980	.1676	1.244	.8466	1.591	1.879	.9133	.5322	2.488	5.978	6.061	.7081	.5995	6.885
.1990	.1688	1.250	.8484	1.603	1.889	.9133	.5294	2.501	6.055	6.137	.7065	.5994	6.856
.2000	.1700	1.257	.8501	1.614	1.899	.9134	.5266	2.513	6.132	6.213	.7049	.5993	6.828
.2010	.1712	1.263	.8519	1.626	1.909	.9135	.5238	2.526	6.211	6.291	.7033	.5992	6.801
.2020	.1724	1.269	.8535	1.638	1.920	.9137	.5210	2.538	6.290	6.369	.7018	.5990	6.774
.2030	.1736	1.276	.8552	1.651	1.930	.9138	.5182	2.551	6.371	6.449	.7002	.5988	6.747
.2040	.1748	1.282	.8570	1.663	1.940	.9139	.5154	2.564	6.452	6.529	.6987	.5987	6.720
.2050	.1760	1.288	.8586	1.675	1.951	.9140	.5127	2.576	6.535	6.611	.6971	.5986	6.694
.2060	.1772	1.294	.8602	1.687	1.961	.9141	.5099	2.589	6.619	6.694	.6956	.5984	6.669
.2070	.1784	1.301	.8619	1.700	1.972	.9142	.5071	2.601	6.703	6.777	.6941	.5982	6.644
.2080	.1796	1.307	.8635	1.712	1.983	.9144	.5044	2.614	6.789	6.862	.6925	.5980	6.619
.2090	.1808	1.313	.8651	1.725	1.994	.9146	.5016	2.626	6.876	6.948	.6910	.5978	6.594
.2100	.1820	1.320	.8667	1.737	2.004	.9147	.4989	2.639	6.963	7.035	.6895	.5976	6.570
.2110	.1832	1.326	.8682	1.750	2.015	.9149	.4962	2.652	7.052	7.123	.6880	.5973	6.547
.2120	.1844	1.332	.8697	1.762	2.026	.9151	.4935	2.664	7.143	7.219	.6865	.5971	6.524
.2130	.1856	1.338	.8713	1.775	2.037	.9153	.4908	2.677	7.234	7.302	.6850	.5969	6.501
.2140	.1868	1.345	.8728	1.788	2.049	.9155	.4881	2.689	7.326	7.394	.6835	.5966	6.479
.2150	.1880	1.351	.8743	1.801	2.060	.9157	.4854	2.702	7.420	7.487	.6821	.5963	6.457
.2160	.1892	1.357	.8757	1.814	2.071	.9159	.4828	2.714	7.514	7.580	.6806	.5960	6.435
.2170	.1904	1.364	.8772	1.827	2.083	.9161	.4801	2.727	7.610	7.675	.6792	.5958	6.413
.2180	.1915	1.370	.8786	1.840	2.094	.9164	.4775	2.739	7.707	7.772	.6777	.5955	6.393
.2190	.1927	1.376	.8801	1.853	2.106	.9166	.4749	2.752	7.805	7.869	.6763	.5952	6.372
.2200	.1939	1.382	.8815	1.867	2.118	.9168	.4722	2.765	7.905	7.968	.6749	.5949	6.351
.2210	.1951	1.389	.8829	1.880	2.129	.9170	.4696	2.777	8.006	8.068	.6735	.5946	6.331
.2220	.1963	1.395	.8842	1.893	2.141	.9173	.4670	2.790	8.108	8.169	.6720	.5943	6.312
.2230	.1975	1.401	.8856	1.907	2.153	.9175	.4644	2.802	8.211	8.272	.6706	.5939	6.292
.2240	.1987	1.407	.8869	1.920	2.165	.9178	.4619	2.815	8.316	8.375	.6692	.5936	6.273
.2250	.1999	1.414	.8883	1.934	2.177	.9181	.4593	2.827	8.422	8.481	.6679	.5933	6.254
.2260	.2011	1.420	.8896	1.948	2.189	.9183	.4567	2.840	8.529	8.587	.6665	.5929	6.236
.2270	.2022	1.426	.8909	1.962	2.202	.9186	.4542	2.853	8.637	8.695	.6651	.5925	6.218
.2280	.2034	1.433	.8922	1.975	2.214	.9189	.4516	2.865	8.756	8.800	.6637	.5921	6.200
.2290	.2046	1.439	.8935	1.989	2.227	.9191	.4491	2.878	8.859	8.915	.6624	.5918	6.182

Table C—2 — Continued

d/L	d/Lo	2πd/L	TANH 2πd/L	SINH 2πd/L	COSH 2πd/L	H/H'o	K	4πd/L	SINH 4πd/L	COSH 4πd/L	n	Cg/Co	M
.2300	.2058	1.445	.8947	2.003	2.239	.9194	.4466	2.890	8.971	9.027	.6611	.5915	6.165
.2310	.2070	1.451	.8960	2.017	2.252	.9197	.4441	2.903	9.085	9.140	.6597	.5911	6.148
.2320	.2082	1.458	.8972	2.032	2.264	.9200	.4416	2.915	9.201	9.255	.6584	.5907	6.131
.2330	.2093	1.464	.8984	2.046	2.277	.9203	.4391	2.928	9.318	9.372	.6571	.5904	6.114
.2340	.2105	1.470	.8996	2.060	2.290	.9206	.4366	2.941	9.437	9.489	.6558	.5900	6.097
.2350	.2117	1.477	.9008	2.075	2.303	.9209	.4342	2.953	9.557	9.609	.6545	.5896	6.081
.2360	.2129	1.483	.9020	2.089	2.316	.9212	.4318	2.966	9.678	9.730	.6532	.5892	6.066
.2370	.2141	1.489	.9032	2.104	2.329	.9215	.4293	2.978	9.801	9.852	.6519	.5888	6.050
.2380	.2152	1.495	.9043	2.118	2.343	.9218	.4269	2.991	9.926	9.976	.6507	.5884	6.034
.2390	.2164	1.502	.9055	2.133	2.356	.9221	.4244	3.003	10.05	10.10	.6494	.5880	6.019
.2400	.2176	1.508	.9066	2.148	2.370	.9225	.4220	3.016	10.18	10.23	.6481	.5876	6.004
.2410	.2188	1.514	.9077	2.163	2.383	.9228	.4196	3.029	10.31	10.36	.6469	.5872	5.990
.2420	.2199	1.521	.9088	2.178	2.397	.9231	.4172	3.041	10.44	10.49	.6456	.5868	5.976
.2430	.2211	1.527	.9099	2.193	2.410	.9234	.4149	3.054	10.57	10.62	.6444	.5863	5.961
.2440	.2223	1.533	.9110	2.208	2.424	.9238	.4125	3.066	10.71	10.75	.6432	.5859	5.947
.2450	.2234	1.539	.9120	2.224	2.438	.9241	.4101	3.079	10.84	10.89	.6420	.5855	5.933
.2460	.2246	1.546	.9131	2.239	2.452	.9244	.4078	3.091	10.98	11.03	.6408	.5851	5.919
.2470	.2258	1.552	.9141	2.255	2.466	.9248	.4055	3.104	11.12	11.17	.6396	.5846	5.906
.2480	.2270	1.558	.9151	2.270	2.480	.9251	.4032	3.116	11.26	11.31	.6384	.5842	5.893
.2490	.2281	1.565	.9162	2.286	2.495	.9255	.4008	3.129	11.40	11.45	.6372	.5838	5.880
.2500	.2293	1.571	.9172	2.301	2.509	.9258	.3985	3.142	11.55	11.59	.6360	.5833	5.867
.2510	.2305	1.577	.9182	2.317	2.524	.9262	.3962	3.154	11.70	11.74	.6348	.5829	5.854
.2520	.2316	1.583	.9191	2.333	2.538	.9265	.3940	3.167	11.84	11.89	.6337	.5824	5.841
.2530	.2328	1.590	.9201	2.349	2.553	.9269	.3917	3.179	11.99	12.04	.6325	.5820	5.829
.2540	.2339	1.596	.9210	2.365	2.568	.9273	.3894	3.192	12.15	12.19	.6314	.5815	5.817
.2550	.2351	1.602	.9220	2.381	2.583	.9276	.3872	3.204	12.30	12.34	.6303	.5811	5.805
.2560	.2363	1.609	.9229	2.398	2.598	.9280	.3849	3.217	12.46	12.50	.6291	.5807	5.793
.2570	.2374	1.615	.9239	2.414	2.613	.9283	.3827	3.230	12.61	12.65	.6280	.5802	5.782
.2580	.2386	1.621	.9248	2.430	2.628	.9287	.3805	3.242	12.77	12.81	.6269	.5797	5.770
.2590	.2398	1.627	.9257	2.447	2.643	.9291	.3783	3.255	12.94	12.98	.6258	.5793	5.759
.2600	.2409	1.634	.9266	2.464	2.659	.9294	.3761	3.267	13.10	13.14	.6247	.5788	5.748
.2610	.2421	1.640	.9275	2.480	2.674	.9298	.3739	3.280	13.27	13.31	.6236	.5784	5.737
.2620	.2432	1.646	.9283	2.497	2.690	.9301	.3717	3.292	13.44	13.47	.6225	.5779	5.726
.2630	.2444	1.653	.9292	2.514	2.706	.9305	.3696	3.305	13.61	13.64	.6215	.5775	5.716
.2640	.2455	1.659	.9301	2.531	2.722	.9309	.3674	3.318	13.78	13.81	.6204	.5770	5.705
.2650	.2467	1.665	.9309	2.548	2.737	.9313	.3653	3.330	13.95	13.99	.6193	.5765	5.695
.2660	.2478	1.671	.9317	2.566	2.754	.9316	.3632	3.343	14.13	14.17	.6183	.5761	5.685
.2670	.2490	1.678	.9326	2.583	2.770	.9320	.3610	3.355	14.31	14.34	.6172	.5756	5.675
.2680	.2501	1.684	.9334	2.600	2.786	.9324	.3589	3.368	14.49	14.53	.6162	.5752	5.665
.2690	.2513	1.690	.9342	2.618	2.803	.9328	.3568	3.380	14.67	14.71	.6152	.5747	5.655
.2700	.2524	1.697	.9350	2.636	2.819	.9331	.3547	3.393	14.86	14.89	.6142	.5742	5.645
.2710	.2536	1.703	.9357	2.653	2.835	.9335	.3527	3.405	15.05	15.08	.6132	.5737	5.636
.2720	.2547	1.709	.9365	2.671	2.852	.9339	.3506	3.418	15.24	15.27	.6122	.5733	5.627
.2730	.2559	1.715	.9373	2.689	2.869	.9343	.3485	3.431	15.43	15.46	.6112	.5728	5.617
.2740	.2570	1.722	.9381	2.707	2.886	.9346	.3465	3.443	15.63	15.66	.6102	.5724	5.608
.2750	.2582	1.728	.9388	2.726	2.903	.9350	.3444	3.456	15.83	15.86	.6092	.5719	5.599
.2760	.2593	1.734	.9396	2.744	2.920	.9354	.3424	3.468	16.03	16.06	.6082	.5714	5.590
.2770	.2605	1.740	.9403	2.762	2.938	.9358	.3404	3.481	16.23	16.26	.6072	.5710	5.582
.2780	.2616	1.747	.9410	2.781	2.955	.9362	.3384	3.493	16.43	16.47	.6063	.5705	5.573
.2790	.2627	1.753	.9417	2.799	2.973	.9366	.3364	3.506	16.64	16.67	.6053	.5701	5.565
.2800	.2639	1.759	.9424	2.818	2.990	.9369	.3344	3.519	16.85	16.88	.6044	.5696	5.556
.2810	.2650	1.766	.9431	2.837	3.008	.9373	.3324	3.531	17.07	17.10	.6035	.5691	5.548
.2820	.2662	1.772	.9438	2.856	3.026	.9377	.3305	3.544	17.28	17.31	.6025	.5687	5.540
.2830	.2673	1.778	.9445	2.875	3.044	.9381	.3285	3.556	17.50	17.53	.6016	.5682	5.532
.2840	.2684	1.784	.9452	2.894	3.062	.9384	.3266	3.569	17.72	17.75	.6007	.5677	5.524
.2850	.2696	1.791	.9458	2.913	3.080	.9388	.3247	3.581	17.95	17.98	.5998	.5673	5.516
.2860	.2707	1.797	.9465	2.933	3.099	.9392	.3227	3.594	18.18	18.20	.5989	.5668	5.509
.2870	.2718	1.803	.9472	2.952	3.117	.9396	.3208	3.607	18.40	18.43	.5980	.5664	5.501
.2880	.2730	1.810	.9478	2.972	3.136	.9400	.3189	3.619	18.64	18.67	.5971	.5659	5.493
.2890	.2741	1.816	.9484	2.992	3.154	.9404	.3170	3.632	18.88	18.90	.5962	.5654	5.486

Table C−2 − Continued

d/L	d/L_o	$\frac{2\pi d}{L}$	$\tanh \frac{2\pi d}{L}$	$\sinh \frac{2\pi d}{L}$	$\cosh \frac{2\pi d}{L}$	H/H'_o	K	$\frac{4\pi d}{L}$	$\sinh \frac{4\pi d}{L}$	$\cosh \frac{4\pi d}{L}$	n	C_G/C_o	M
.2900	.2752	1.822	.9491	3.012	3.173	.9407	.3151	3.644	19.11	19.14	.5953	.5650	5.479
.2910	.2764	1.828	.9497	3.032	3.192	.9411	.3133	3.657	19.36	19.38	.5945	.5645	5.472
.2920	.2775	1.835	.9503	3.052	3.211	.9415	.3114	3.669	19.60	19.63	.5936	.5641	5.465
.2930	.2786	1.841	.9509	3.072	3.231	.9419	.3095	3.682	19.85	19.87	.5927	.5636	5.458
.2940	.2797	1.847	.9515	3.093	3.250	.9422	.3077	3.695	20.10	20.13	.5919	.5632	5.451
.2950	.2809	1.854	.9521	3.113	3.269	.9426	.3059	3.707	20.36	20.38	.5911	.5627	5.444
.2960	.2820	1.860	.9527	3.133	3.289	.9430	.3040	3.720	20.61	20.64	.5902	.5622	5.437
.2970	.2831	1.866	.9532	3.154	3.309	.9434	.3022	3.732	20.87	20.90	.5894	.5618	5.431
.2980	.2842	1.872	.9538	3.175	3.329	.9437	.3004	3.745	21.14	21.16	.5886	.5614	5.424
.2990	.2854	1.879	.9544	3.196	3.349	.9441	.2986	3.757	21.41	21.43	.5878	.5610	5.418
.3000	.2865	1.885	.9549	3.217	3.369	.9445	.2968	3.770	21.68	21.70	.5870	.5605	5.412
.3010	.2876	1.891	.9555	3.238	3.389	.9449	.2951	3.782	21.95	21.97	.5862	.5601	5.405
.3020	.2887	1.898	.9560	3.260	3.410	.9452	.2933	3.795	22.23	22.25	.5854	.5596	5.399
.3030	.2898	1.904	.9566	3.281	3.430	.9456	.2915	3.808	22.51	22.53	.5846	.5592	5.393
.3040	.2910	1.910	.9571	3.303	3.451	.9459	.2898	3.820	22.80	22.82	.5838	.5587	5.387
.3050	.2921	1.916	.9576	3.325	3.472	.9463	.2880	3.833	23.08	23.11	.5830	.5583	5.381
.3060	.2932	1.923	.9581	3.347	3.493	.9467	.2863	3.845	23.38	23.40	.5823	.5579	5.376
.3070	.2943	1.929	.9586	3.368	3.514	.9471	.2846	3.858	23.67	23.69	.5815	.5574	5.370
.3080	.2954	1.935	.9592	3.391	3.535	.9474	.2829	3.870	23.97	23.99	.5807	.5570	5.364
.3090	.2965	1.942	.9597	3.413	3.556	.9478	.2812	3.883	24.28	24.30	.5800	.5566	5.359
.3100	.2977	1.948	.9602	3.435	3.578	.9482	.2795	3.896	24.58	24.60	.5792	.5562	5.353
.3110	.2988	1.954	.9606	3.458	3.600	.9485	.2778	3.908	24.89	24.91	.5785	.5557	5.348
.3120	.2999	1.960	.9611	3.481	3.621	.9489	.2761	3.921	25.21	25.23	.5778	.5553	5.342
.3130	.3010	1.967	.9616	3.503	3.643	.9493	.2745	3.933	25.53	25.55	.5770	.5549	5.337
.3140	.3021	1.973	.9621	3.526	3.665	.9496	.2728	3.946	25.85	25.87	.5763	.5545	5.332
.3150	.3032	1.979	.9625	3.549	3.688	.9500	.2712	3.958	26.18	26.20	.5756	.5540	5.327
.3160	.3043	1.986	.9630	3.573	3.710	.9504	.2695	3.971	26.51	26.53	.5749	.5536	5.321
.3170	.3054	1.992	.9634	3.596	3.733	.9508	.2679	3.984	26.84	26.86	.5742	.5532	5.316
.3180	.3065	1.998	.9639	3.620	3.755	.9511	.2663	3.996	27.18	27.20	.5735	.5528	5.311
.3190	.3076	2.004	.9643	3.643	3.778	.9514	.2647	4.009	27.53	27.55	.5728	.5524	5.307
.3200	.3087	2.011	.9648	3.667	3.801	.9518	.2631	4.021	27.88	27.89	.5721	.5520	5.302
.3210	.3098	2.017	.9652	3.691	3.824	.9521	.2615	4.034	28.23	28.25	.5714	.5516	5.297
.3220	.3109	2.023	.9656	3.715	3.847	.9525	.2599	4.046	28.59	28.60	.5708	.5512	5.292
.3230	.3120	2.030	.9661	3.739	3.871	.9528	.2583	4.059	28.95	28.97	.5701	.5508	5.288
.3240	.3131	2.036	.9665	3.764	3.894	.9532	.2568	4.072	29.31	29.33	.5694	.5504	5.283
.3250	.3142	2.042	.9669	3.788	3.918	.9535	.2552	4.084	29.69	29.70	.5688	.5500	5.279
.3260	.3153	2.048	.9673	3.813	3.942	.9539	.2537	4.097	30.06	30.08	.5681	.5496	5.274
.3270	.3164	2.055	.9677	3.838	3.966	.9542	.2521	4.109	30.44	30.46	.5675	.5492	5.270
.3280	.3175	2.061	.9681	3.863	3.990	.9545	.2506	4.122	30.83	30.84	.5669	.5488	5.266
.3290	.3186	2.067	.9685	3.888	4.015	.9549	.2491	4.134	31.22	31.23	.5662	.5484	5.261
.3300	.3197	2.074	.9689	3.913	4.039	.9552	.2476	4.147	31.61	31.63	.5656	.5480	5.257
.3310	.3208	2.080	.9692	3.939	4.064	.9555	.2461	4.159	32.01	32.03	.5650	.5476	5.253
.3320	.3219	2.086	.9696	3.964	4.088	.9559	.2446	4.172	32.42	32.43	.5644	.5472	5.249
.3330	.3230	2.092	.9700	3.990	4.114	.9562	.2431	4.185	32.83	32.84	.5637	.5468	5.245
.3340	.3241	2.099	.9704	4.016	4.139	.9566	.2416	4.197	33.24	33.26	.5631	.5464	5.241
.3350	.3252	2.105	.9707	4.042	4.164	.9569	.2402	4.210	33.66	33.68	.5625	.5461	5.237
.3360	.3263	2.111	.9711	4.069	4.189	.9572	.2387	4.222	34.09	34.10	.5619	.5457	5.233
.3370	.3274	2.117	.9715	4.095	4.215	.9576	.2373	4.235	34.52	34.53	.5613	.5453	5.229
.3380	.3285	2.124	.9718	4.121	4.241	.9579	.2358	4.247	34.96	34.97	.5608	.5449	5.225
.3390	.3296	2.130	.9722	4.148	4.267	.9582	.2344	4.260	35.40	35.41	.5602	.5446	5.222
.3400	.3307	2.136	.9725	4.175	4.293	.9585	.2329	4.273	35.85	35.86	.5596	.5442	5.218
.3410	.3317	2.143	.9728	4.202	4.319	.9589	.2315	4.285	36.30	36.31	.5590	.5438	5.214
.3420	.3328	2.149	.9732	4.229	4.346	.9592	.2301	4.298	36.76	36.77	.5585	.5435	5.211
.3430	.3339	2.155	.9735	4.256	4.372	.9595	.2287	4.310	37.22	37.24	.5579	.5431	5.207
.3440	.3350	2.161	.9738	4.284	4.399	.9598	.2273	4.323	37.70	37.71	.5573	.5427	5.204
.3450	.3361	2.168	.9742	4.312	4.426	.9601	.2259	4.335	38.17	38.19	.5568	.5424	5.200
.3460	.3372	2.174	.9745	4.340	4.454	.9604	.2245	4.348	38.65	38.67	.5562	.5420	5.197
.3470	.3383	2.180	.9748	4.368	4.481	.9608	.2232	4.361	39.14	39.16	.5557	.5417	5.193
.3480	.3393	2.187	.9751	4.396	4.509	.9611	.2218	4.373	39.64	39.65	.5552	.5413	5.190
.3490	.3404	2.193	.9754	4.424	4.536	.9614	.2205	4.386	40.14	40.15	.5546	.5410	5.187

C-24

Table C–2 – Continued

d/L	d/L_0	$\frac{2\pi d}{L}$	$\tanh\frac{2\pi d}{L}$	$\sinh\frac{2\pi d}{L}$	$\cosh\frac{2\pi d}{L}$	H/H_0'	K	$\frac{4\pi d}{L}$	$\sinh\frac{4\pi d}{L}$	$\cosh\frac{4\pi d}{L}$	n	C_G/C_0	M
.3500	.3415	2.199	.9757	4.453	4.564	.9617	.2191	4.398	40.65	40.66	.5541	.5406	5.184
.3510	.3426	2.205	.9760	4.482	4.592	.9620	.2178	4.411	41.16	41.17	.5536	.5403	5.181
.3520	.3437	2.212	.9763	4.511	4.620	.9623	.2164	4.423	41.68	41.70	.5531	.5400	5.177
.3530	.3447	2.218	.9766	4.540	4.649	.9626	.2151	4.436	42.21	42.22	.5525	.5396	5.174
.3540	.3458	2.224	.9769	4.569	4.678	.9629	.2138	4.449	42.74	42.76	.5520	.5393	5.171
.3550	.3469	2.231	.9772	4.600	4.706	.9632	.2125	4.461	43.28	43.30	.5515	.5389	5.168
.3560	.3480	2.237	.9774	4.628	4.735	.9635	.2112	4.474	43.83	43.84	.5510	.5386	5.165
.3570	.3491	2.243	.9777	4.658	4.764	.9638	.2099	4.486	44.39	44.40	.5505	.5383	5.162
.3580	.3501	2.249	.9780	4.688	4.794	.9641	.2086	4.499	44.95	44.96	.5500	.5379	5.159
.3590	.3512	2.256	.9783	4.719	4.823	.9644	.2073	4.511	45.52	45.53	.5496	.5376	5.156
.3600	.3523	2.262	.9785	4.749	4.853	.9647	.2060	4.524	46.09	46.10	.5491	.5373	5.154
.3610	.3534	2.268	.9788	4.779	4.883	.9650	.2048	4.536	46.68	46.69	.5486	.5370	5.151
.3620	.3544	2.275	.9791	4.810	4.913	.9652	.2035	4.549	47.27	47.28	.5481	.5367	5.148
.3630	.3555	2.281	.9793	4.840	4.943	.9655	.2023	4.562	47.86	47.87	.5477	.5363	5.145
.3640	.3566	2.287	.9796	4.872	4.974	.9658	.2010	4.574	48.47	48.48	.5472	.5360	5.143
.3650	.3576	2.293	.9798	4.904	5.005	.9661	.1998	4.587	49.08	49.09	.5467	.5357	5.140
.3660	.3587	2.300	.9801	4.935	5.035	.9664	.1986	4.599	49.70	49.71	.5463	.5354	5.137
.3670	.3598	2.306	.9803	4.967	5.067	.9667	.1974	4.612	50.33	50.34	.5458	.5351	5.135
.3680	.3609	2.312	.9806	4.999	5.098	.9670	.1962	4.624	50.97	50.98	.5454	.5348	5.132
.3690	.3619	2.319	.9808	5.031	5.129	.9672	.1950	4.637	51.61	51.62	.5449	.5345	5.130
.3700	.3630	2.325	.9811	5.063	5.161	.9675	.1938	4.650	52.27	52.28	.5445	.5342	5.127
.3710	.3641	2.331	.9813	5.096	5.193	.9678	.1926	4.662	52.93	52.94	.5440	.5339	5.125
.3720	.3651	2.337	.9815	5.129	5.225	.9680	.1914	4.675	53.60	53.61	.5436	.5336	5.122
.3730	.3662	2.346	.9817	5.161	5.257	.9683	.1902	4.687	54.27	54.28	.5432	.5333	5.120
.3740	.3673	2.350	.9820	5.195	5.290	.9686	.1890	4.700	54.99	54.97	.5427	.5330	5.118
.3750	.3683	2.356	.9822	5.228	5.322	.9688	.1879	4.712	55.66	55.66	.5423	.5327	5.115
.3760	.3694	2.363	.9824	5.262	5.356	.9691	.1867	4.725	56.36	56.37	.5419	.5324	5.113
.3770	.3705	2.369	.9826	5.295	5.389	.9694	.1856	4.738	57.07	57.08	.5415	.5321	5.111
.3780	.3715	2.375	.9829	5.329	5.422	.9696	.1844	4.750	57.79	57.80	.5411	.5318	5.109
.3790	.3726	2.381	.9831	5.363	5.456	.9699	.1833	4.763	58.53	58.53	.5407	.5315	5.106
.3800	.3736	2.388	.9833	5.398	5.490	.9702	.1822	4.775	59.27	59.27	.5403	.5313	5.104
.3810	.3747	2.394	.9835	5.432	5.524	.9704	.1810	4.788	60.01	60.02	.5399	.5310	5.102
.3820	.3758	2.400	.9837	5.467	5.558	.9707	.1799	4.800	60.77	60.78	.5395	.5307	5.100
.3830	.3768	2.407	.9839	5.502	5.593	.9709	.1788	4.813	61.54	61.55	.5391	.5304	5.098
.3840	.3779	2.413	.9841	5.537	5.627	.9712	.1777	4.826	62.32	62.33	.5387	.5301	5.096
.3850	.3790	2.419	.9843	5.573	5.662	.9714	.1766	4.838	63.11	63.12	.5383	.5299	5.094
.3860	.3800	2.425	.9845	5.609	5.697	.9717	.1755	4.851	63.91	63.91	.5380	.5296	5.092
.3870	.3811	2.432	.9847	5.645	5.732	.9719	.1744	4.863	64.72	64.72	.5376	.5293	5.090
.3880	.3821	2.438	.9849	5.681	5.768	.9721	.1734	4.876	65.53	65.54	.5372	.5291	5.088
.3890	.3832	2.444	.9850	5.717	5.804	.9724	.1723	4.889	66.40	66.40	.5368	.5288	5.086
.3900	.3842	2.450	.9852	5.753	5.840	.9726	.1712	4.901	67.20	67.21	.5365	.5285	5.084
.3910	.3853	2.457	.9854	5.790	5.876	.9729	.1702	4.913	68.05	68.06	.5361	.5283	5.082
.3920	.3864	2.463	.9856	5.827	5.913	.9731	.1691	4.926	68.91	68.92	.5357	.5280	5.080
.3930	.3874	2.469	.9858	5.865	5.949	.9733	.1681	4.939	69.78	69.79	.5354	.5278	5.078
.3940	.3885	2.476	.9860	5.902	5.988	.9736	.1670	4.951	70.67	70.67	.5350	.5275	5.077
.3950	.3895	2.482	.9861	5.940	6.024	.9738	.1660	4.964	71.56	71.57	.5347	.5273	5.075
.3960	.3906	2.488	.9863	5.978	6.061	.9740	.1650	4.976	72.47	72.47	.5343	.5270	5.073
.3970	.3916	2.494	.9865	6.016	6.099	.9743	.1640	4.989	73.38	73.39	.5340	.5268	5.071
.3980	.3927	2.501	.9866	6.054	6.137	.9745	.1630	5.001	74.31	74.32	.5337	.5265	5.070
.3990	.3937	2.507	.9868	6.093	6.175	.9747	.1619	5.014	75.25	75.26	.5333	.5263	5.068
.4000	.3948	2.513	.9870	6.132	6.213	.9749	.1609	5.027	76.20	76.21	.5330	.5260	5.066
.4010	.3958	2.520	.9871	6.172	6.252	.9752	.1600	5.039	77.16	77.17	.5327	.5258	5.064
.4020	.3969	2.526	.9873	6.210	6.290	.9754	.1590	5.052	78.14	78.15	.5323	.5256	5.063
.4030	.3979	2.532	.9874	6.250	6.330	.9756	.1580	5.064	79.13	79.14	.5320	.5253	5.061
.4040	.3990	2.538	.9876	6.290	6.369	.9758	.1570	5.077	80.13	80.14	.5317	.5251	5.060
.4050	.4000	2.545	.9878	6.330	6.409	.9760	.1560	5.089	81.14	81.15	.5314	.5249	5.058
.4060	.4011	2.551	.9879	6.371	6.449	.9763	.1551	5.102	82.17	82.18	.5310	.5246	5.056
.4070	.4021	2.557	.9881	6.412	6.489	.9765	.1541	5.115	83.21	83.21	.5307	.5244	5.055
.4080	.4032	2.564	.9882	6.452	6.529	.9767	.1532	5.127	84.25	84.26	.5304	.5242	5.053
.4090	.4042	2.570	.9883	6.493	6.571	.9769	.1522	5.140	85.33	85.33	.5301	.5239	5.052

Table C-2 — Continued

d/L	d/L_0	$\frac{2\pi d}{L}$	$\tanh \frac{2\pi d}{L}$	$\sinh \frac{2\pi d}{L}$	$\cosh \frac{2\pi d}{L}$	H/H_0'	K	$\frac{4\pi d}{L}$	$\sinh \frac{4\pi d}{L}$	$\cosh \frac{4\pi d}{L}$	n	C_G/C_0	M
.4100	.4053	2.576	.9885	6.535	6.611	.9771	.1513	5.152	86.41	86.41	.5298	.5237	5.050
.4110	.4063	2.582	.9886	6.577	6.653	.9773	.1503	5.165	87.50	87.50	.5295	.5235	5.049
.4120	.4074	2.589	.9888	6.619	6.694	.9775	.1494	5.177	88.61	88.61	.5292	.5233	5.048
.4130	.4084	2.595	.9889	6.661	6.736	.9777	.1485	5.190	89.73	89.73	.5289	.5231	5.046
.4140	.4095	2.601	.9891	6.703	6.777	.9779	.1476	5.202	90.87	90.87	.5286	.5228	5.045
.4150	.4105	2.608	.9892	6.746	6.819	.9781	.1466	5.215	92.02	92.02	.5283	.5226	5.043
.4160	.4116	2.614	.9893	6.789	6.862	.9783	.1457	5.228	93.18	93.18	.5281	.5224	5.042
.4170	.4126	2.620	.9895	6.832	6.905	.9785	.1448	5.240	94.36	94.36	.5278	.5222	5.041
.4180	.4136	2.626	.9896	6.876	6.948	.9787	.1439	5.253	95.55	95.55	.5275	.5220	5.039
.4190	.4147	2.633	.9897	6.920	6.992	.9789	.1430	5.265	96.76	96.76	.5272	.5218	5.038
.4200	.4157	2.639	.9899	6.963	7.035	.9791	.1422	5.278	97.98	97.98	.5269	.5216	5.037
.4210	.4168	2.645	.9900	7.008	7.079	.9793	.1413	5.290	99.22	99.22	.5267	.5214	5.035
.4220	.4178	2.652	.9901	7.052	7.123	.9795	.1404	5.303	100.5	100.5	.5264	.5212	5.034
.4230	.4189	2.658	.9902	7.097	7.167	.9797	.1395	5.316	101.7	101.7	.5261	.5210	5.033
.4240	.4199	2.664	.9903	7.143	7.212	.9799	.1387	5.328	103.0	103.0	.5259	.5208	5.032
.4250	.4210	2.670	.9905	7.188	7.257	.9801	.1378	5.341	104.3	104.3	.5256	.5206	5.030
.4260	.4220	2.677	.9906	7.233	7.302	.9803	.1370	5.353	105.7	105.7	.5253	.5204	5.029
.4270	.4230	2.683	.9907	7.280	7.348	.9804	.1361	5.366	107.0	107.0	.5251	.5202	5.028
.4280	.4241	2.689	.9908	7.326	7.394	.9806	.1352	5.378	108.3	108.3	.5248	.5200	5.027
.4290	.4251	2.696	.9909	7.373	7.440	.9808	.1344	5.391	109.7	109.7	.5246	.5198	5.026
.4300	.4262	2.702	.9910	7.420	7.487	.9810	.1336	5.404	111.1	111.1	.5243	.5196	5.025
.4310	.4272	2.708	.9912	7.467	7.534	.9811	.1327	5.416	112.5	112.5	.5241	.5194	5.023
.4320	.4282	2.714	.9913	7.514	7.580	.9813	.1319	5.429	113.9	113.9	.5238	.5193	5.022
.4330	.4293	2.721	.9914	7.562	7.628	.9815	.1311	5.441	115.4	115.4	.5236	.5191	5.021
.4340	.4303	2.727	.9915	7.610	7.673	.9817	.1303	5.454	116.8	116.8	.5233	.5189	5.020
.4350	.4313	2.733	.9916	7.659	7.723	.9818	.1295	5.466	118.3	118.3	.5231	.5187	5.019
.4360	.4324	2.740	.9917	7.707	7.772	.9820	.1287	5.479	119.8	119.8	.5229	.5185	5.018
.4370	.4334	2.746	.9918	7.756	7.821	.9822	.1279	5.492	121.3	121.3	.5226	.5183	5.017
.4380	.4345	2.752	.9919	7.805	7.869	.9823	.1271	5.504	122.8	122.8	.5224	.5182	5.016
.4390	.4355	2.758	.9920	7.855	7.918	.9825	.1263	5.517	124.4	124.4	.5222	.5180	5.015
.4400	.4365	2.765	.9921	7.905	7.968	.9827	.1255	5.529	126.0	126.0	.5219	.5178	5.014
.4410	.4376	2.771	.9922	7.955	8.018	.9828	.1247	5.542	127.6	127.6	.5217	.5177	5.013
.4420	.4386	2.777	.9923	8.006	8.068	.9830	.1239	5.554	129.2	129.2	.5215	.5175	5.012
.4430	.4396	2.784	.9924	8.057	8.119	.9831	.1232	5.567	130.8	130.8	.5213	.5173	5.011
.4440	.4407	2.790	.9925	8.107	8.169	.9833	.1224	5.579	132.6	132.6	.5210	.5171	5.010
.4450	.4417	2.796	.9926	8.159	8.220	.9835	.1217	5.592	134.1	134.1	.5208	.5170	5.009
.4460	.4427	2.802	.9927	8.211	8.272	.9836	.1209	5.605	135.8	135.8	.5206	.5168	5.008
.4470	.4438	2.809	.9928	8.263	8.322	.9838	.1202	5.617	137.6	137.6	.5204	.5166	5.007
.4480	.4448	2.815	.9929	8.316	8.376	.9839	.1194	5.630	139.3	139.3	.5202	.5165	5.006
.4490	.4458	2.821	.9929	8.369	8.428	.9841	.1186	5.642	141.1	141.1	.5200	.5163	5.005
.4500	.4469	2.827	.9930	8.421	8.480	.9842	.1179	5.655	142.8	142.8	.5198	.5162	5.004
.4510	.4479	2.834	.9931	8.475	8.534	.9844	.1172	5.667	144.7	144.7	.5196	.5160	5.003
.4520	.4489	2.840	.9932	8.529	8.587	.9845	.1165	5.680	146.5	146.5	.5194	.5159	5.003
.4530	.4500	2.846	.9933	8.583	8.641	.9847	.1157	5.693	148.3	148.3	.5192	.5157	5.002
.4540	.4510	2.853	.9934	8.638	8.695	.9848	.1150	5.705	150.2	150.2	.5190	.5156	5.001
.4550	.4520	2.859	.9935	8.692	8.750	.9850	.1143	5.718	152.1	152.1	.5188	.5154	5.000
.4560	.4531	2.865	.9935	8.747	8.804	.9851	.1136	5.730	154.0	154.0	.5186	.5153	4.999
.4570	.4541	2.871	.9936	8.803	8.859	.9852	.1129	5.743	156.0	156.0	.5184	.5151	4.999
.4580	.4551	2.878	.9937	8.859	8.915	.9854	.1122	5.755	158.0	158.0	.5182	.5150	4.998
.4590	.4561	2.884	.9938	8.915	8.971	.9855	.1115	5.768	159.9	159.9	.5180	.5148	4.997
.4600	.4572	2.890	.9938	8.972	9.022	.9857	.1108	5.781	162.0	162.0	.5178	.5147	4.996
.4610	.4582	2.897	.9939	9.029	9.084	.9858	.1101	5.793	164.0	164.0	.5177	.5145	4.995
.4620	.4592	2.903	.9940	9.085	9.140	.9859	.1094	5.806	166.1	166.1	.5175	.5144	4.995
.4630	.4603	2.909	.9941	9.143	9.197	.9861	.1087	5.818	168.2	168.2	.5173	.5142	4.994
.4640	.4613	2.915	.9941	9.201	9.255	.9862	.1080	5.831	170.3	170.3	.5171	.5141	4.993
.4650	.4623	2.922	.9942	9.260	9.313	.9863	.1074	5.843	172.5	172.5	.5169	.5140	4.992
.4660	.4633	2.928	.9943	9.318	9.372	.9865	.1067	5.856	174.7	174.7	.5168	.5138	4.992
.4670	.4644	2.934	.9944	9.378	9.431	.9866	.1060	5.869	176.9	176.9	.5166	.5137	4.991
.4680	.4654	2.941	.9944	9.436	9.489	.9867	.1054	5.881	179.1	179.1	.5164	.5136	4.990
.4690	.4664	2.947	.9945	9.496	9.549	.9868	.1047	5.894	181.4	181.4	.5162	.5134	4.990

Table C-2 — Continued

d/L	d/L$_o$	$\frac{2\pi d}{L}$	tanh $\frac{2\pi d}{L}$	sinh $\frac{2\pi d}{L}$	cosh $\frac{2\pi d}{L}$	H/H$_o$'	K	$\frac{4\pi d}{L}$	sinh $\frac{4\pi d}{L}$	cosh $\frac{4\pi d}{L}$	n	C$_G$/C$_o$	M
.4700	.4675	2.953	.9946	9.557	9.609	.9870	.1041	5.906	183.7	183.7	.5161	.5133	4.989
.4710	.4685	2.959	.9946	9.617	9.669	.9871	.1034	5.919	186.0	186.0	.5159	.5131	4.988
.4720	.4695	2.966	.9947	9.678	9.730	.9872	.1028	5.931	188.3	188.3	.5157	.5130	4.987
.4730	.4705	2.972	.9948	9.740	9.791	.9873	.1021	5.944	190.7	190.7	.5156	.5129	4.987
.4740	.4716	2.978	.9948	9.801	9.852	.9875	.1015	5.956	193.1	193.1	.5154	.5128	4.986
.4750	.4726	2.985	.9949	9.863	9.914	.9876	.1009	5.969	195.6	195.6	.5153	.5126	4.986
.4760	.4736	2.991	.9950	9.926	9.976	.9877	.1002	5.982	198.0	198.0	.5151	.5125	4.985
.4770	.4746	2.997	.9950	9.989	10.04	.9878	.09961	5.994	200.5	200.5	.5149	.5124	4.984
.4780	.4757	3.003	.9951	10.05	10.10	.9880	.09899	6.007	203.1	203.1	.5148	.5123	4.984
.4790	.4767	3.010	.9952	10.12	10.17	.9881	.09838	6.019	205.6	205.6	.5146	.5121	4.983
.4800	.4777	3.016	.9952	10.18	10.23	.9882	.09776	6.032	208.2	208.2	.5145	.5120	4.983
.4810	.4787	3.022	.9953	10.24	10.29	.9883	.09715	6.044	210.9	210.9	.5143	.5119	4.982
.4820	.4798	3.029	.9953	10.31	10.36	.9884	.09655	6.057	213.5	213.5	.5142	.5118	4.981
.4830	.4808	3.035	.9954	10.37	10.42	.9885	.09595	6.070	216.2	216.2	.5140	.5117	4.981
.4840	.4818	3.041	.9954	10.44	10.49	.9887	.09535	6.082	219.0	219.0	.5139	.5115	4.980
.4850	.4828	3.047	.9955	10.51	10.55	.9888	.09475	6.095	221.7	221.7	.5137	.5114	4.980
.4860	.4838	3.054	.9956	10.57	10.62	.9889	.09416	6.107	224.5	224.5	.5136	.5113	4.979
.4870	.4849	3.060	.9956	10.64	10.69	.9890	.09358	6.120	227.4	227.4	.5135	.5112	4.978
.4880	.4859	3.066	.9957	10.71	10.75	.9891	.09300	6.132	230.3	230.3	.5133	.5111	4.978
.4890	.4869	3.073	.9957	10.77	10.82	.9892	.09241	6.145	233.2	233.2	.5132	.5110	4.977
.4900	.4879	3.079	.9958	10.84	10.89	.9893	.09183	6.158	236.1	236.1	.5130	.5109	4.977
.4910	.4890	3.085	.9958	10.91	10.96	.9894	.09126	6.170	239.1	239.1	.5129	.5108	4.976
.4920	.4900	3.091	.9959	10.98	11.03	.9895	.09069	6.183	242.1	242.1	.5128	.5107	4.976
.4930	.4910	3.098	.9959	11.05	11.10	.9896	.09013	6.195	245.2	245.2	.5126	.5106	4.975
.4940	.4920	3.104	.9960	11.12	11.17	.9897	.08957	6.208	248.3	248.3	.5125	.5104	4.975
.4950	.4930	3.110	.9960	11.19	11.24	.9898	.08901	6.220	251.4	251.4	.5124	.5103	4.974
.4960	.4941	3.117	.9961	11.26	11.31	.9899	.08845	6.233	254.6	254.6	.5122	.5102	4.974
.4970	.4951	3.123	.9961	11.33	11.38	.9900	.08790	6.246	257.8	257.8	.5121	.5101	4.973
.4980	.4961	3.129	.9962	11.40	11.45	.9901	.08736	6.258	261.1	261.1	.5120	.5100	4.973
.4990	.4971	3.135	.9962	11.48	11.52	.9902	.08681	6.271	264.4	264.4	.5119	.5099	4.972
.5000	.4981	3.142	.9963	11.55	11.59	.9903	.08627	6.283	267.7	267.7	.5117	.5098	4.972
.5010	.4992	3.148	.9963	11.62	11.67	.9904	.08573	6.296	271.1	271.1	.5116	.5097	4.971
.5020	.5002	3.154	.9964	11.70	11.74	.9905	.08519	6.308	274.5	274.5	.5115	.5096	4.971
.5030	.5012	3.160	.9964	11.77	11.81	.9906	.08466	6.321	278.0	278.0	.5114	.5095	4.971
.5040	.5022	3.167	.9965	11.84	11.89	.9907	.08413	6.333	281.5	281.5	.5112	.5094	4.970
.5050	.5032	3.173	.9965	11.92	11.96	.9908	.08361	6.346	285.1	285.1	.5111	.5093	4.970
.5060	.5043	3.179	.9965	11.99	12.03	.9909	.08309	6.359	288.7	288.7	.5110	.5092	4.969
.5070	.5053	3.186	.9966	12.07	12.11	.9910	.08257	6.371	292.4	292.4	.5109	.5092	4.969
.5080	.5063	3.192	.9966	12.15	12.19	.9911	.08205	6.384	296.1	296.1	.5108	.5091	4.968
.5090	.5073	3.198	.9967	12.22	12.26	.9911	.08154	6.396	299.8	299.8	.5107	.5090	4.968
.5100	.5083	3.204	.9967	12.30	12.34	.9912	.08103	6.409	303.6	303.6	.5106	.5089	4.967
.5110	.5093	3.211	.9968	12.38	12.42	.9913	.08053	6.421	307.4	307.4	.5104	.5088	4.967
.5120	.5104	3.217	.9968	12.46	12.50	.9914	.08002	6.434	311.3	311.3	.5103	.5087	4.967
.5130	.5114	3.223	.9968	12.53	12.57	.9915	.07952	6.447	315.4	315.4	.5102	.5086	4.966
.5140	.5124	3.230	.9969	12.62	12.65	.9916	.07903	6.459	319.2	319.2	.5101	.5085	4.966
.5150	.5134	3.236	.9969	12.70	12.74	.9917	.07853	6.472	323.3	323.3	.5100	.5084	4.965
.5160	.5144	3.242	.9970	12.77	12.81	.9917	.07804	6.484	327.4	327.4	.5099	.5084	4.965
.5170	.5154	3.248	.9970	12.86	12.89	.9918	.07756	6.497	331.5	331.5	.5098	.5083	4.965
.5180	.5165	3.255	.9970	12.94	12.98	.9919	.07707	6.509	335.7	335.7	.5097	.5082	4.964
.5190	.5175	3.261	.9971	13.02	13.06	.9920	.07659	6.522	339.9	339.9	.5096	.5081	4.964
.5200	.5185	3.267	.9971	13.10	13.14	.9921	.07611	6.535	344.2	344.2	.5095	.5080	4.964
.5210	.5195	3.274	.9971	13.18	13.22	.9921	.07564	6.547	348.2	348.2	.5094	.5079	4.963
.5220	.5205	3.280	.9972	13.27	13.30	.9922	.07517	6.560	353.0	353.0	.5093	.5079	4.963
.5230	.5215	3.286	.9972	13.35	13.38	.9923	.07469	6.572	357.5	357.5	.5092	.5078	4.963
.5240	.5226	3.292	.9972	13.44	13.47	.9924	.07422	6.585	362.0	362.0	.5091	.5077	4.962
.5250	.5236	3.299	.9973	13.52	13.56	.9925	.07376	6.597	366.6	366.6	.5090	.5076	4.962
.5260	.5246	3.305	.9973	13.61	13.64	.9925	.07330	6.610	371.2	371.2	.5089	.5076	4.962
.5270	.5256	3.311	.9974	13.69	13.72	.9926	.07284	6.622	375.9	375.9	.5088	.5075	4.961
.5280	.5266	3.318	.9974	13.78	13.81	.9927	.07239	6.635	380.3	380.3	.5087	.5074	4.961
.5290	.5276	3.324	.9974	13.86	13.89	.9928	.07194	6.648	385.5	385.5	.5086	.5073	4.961

Table C-2 – Continued

d/L	d/L_o	$\frac{2\pi d}{L}$	$\tanh\frac{2\pi d}{L}$	$\sinh\frac{2\pi d}{L}$	$\cosh\frac{2\pi d}{L}$	H/H_o'	K	$\frac{4\pi d}{L}$	$\sinh\frac{4\pi d}{L}$	$\cosh\frac{4\pi d}{L}$	n	C_G/C_o	M
.5300	.5286	3.330	.9974	13.95	13.99	.9929	.07149	6.600	390.3	390.3	.5085	.5072	4.960
.5310	.5297	3.336	.9975	14.04	14.08	.9929	.07104	6.673	395.3	395.3	.5084	.5072	4.960
.5320	.5307	3.343	.9975	14.13	14.17	.9930	.07059	6.685	400.3	400.3	.5084	.5071	4.960
.5330	.5317	3.349	.9975	14.22	14.25	.9931	.07016	6.698	405.3	405.3	.5083	.5070	4.959
.5340	.5327	3.355	.9976	14.31	14.34	.9931	.06972	6.710	410.5	410.5	.5082	.5069	4.959
.5350	.5337	3.362	.9976	14.40	14.43	.9932	.06928	6.723	415.6	415.6	.5081	.5069	4.959
.5360	.5347	3.368	.9976	14.49	14.52	.9933	.06885	6.736	420.9	420.9	.5080	.5068	4.958
.5370	.5357	3.374	.9977	14.58	14.62	.9933	.06842	6.748	426.2	426.2	.5079	.5067	4.958
.5380	.5368	3.380	.9977	14.67	14.71	.9934	.06799	6.761	431.6	431.6	.5078	.5067	4.958
.5390	.5378	3.387	.9977	14.77	14.80	.9935	.06757	6.773	437.1	437.1	.5077	.5066	4.958
.5400	.5388	3.393	.9977	14.86	14.89	.9935	.06715	6.786	442.6	442.6	.5077	.5065	4.957
.5410	.5398	3.399	.9978	14.95	14.99	.9936	.06673	6.798	448.2	448.2	.5076	.5065	4.957
.5420	.5408	3.405	.9978	15.05	15.08	.9937	.06631	6.811	453.9	453.9	.5075	.5064	4.957
.5430	.5418	3.412	.9978	15.14	15.18	.9937	.06589	6.824	459.6	459.6	.5074	.5063	4.956
.5440	.5428	3.418	.9979	15.25	15.27	.9938	.06548	6.836	465.4	465.4	.5073	.5063	4.956
.5450	.5438	3.424	.9979	15.34	15.37	.9939	.06507	6.849	471.2	471.2	.5073	.5062	4.956
.5460	.5449	3.431	.9979	15.43	15.46	.9939	.06467	6.861	477.2	477.2	.5072	.5061	4.956
.5470	.5459	3.437	.9979	15.53	15.56	.9940	.06426	6.874	483.3	483.3	.5071	.5061	4.955
.5480	.5469	3.443	.9980	15.63	15.66	.9941	.06386	6.886	489.4	489.4	.5070	.5060	4.955
.5490	.5479	3.449	.9980	15.73	15.76	.9941	.06346	6.899	495.6	495.6	.5070	.5059	4.955
.5500	.5489	3.456	.9980	15.83	15.86	.9942	.06306	6.912	501.9	501.9	.5069	.5059	4.955
.5510	.5499	3.462	.9980	15.93	15.96	.9942	.06267	6.924	508.2	508.2	.5068	.5058	4.954
.5520	.5509	3.468	.9981	16.03	16.06	.9943	.06228	6.937	514.6	514.6	.5067	.5058	4.954
.5530	.5519	3.475	.9981	16.13	16.16	.9944	.06189	6.949	521.1	521.1	.5067	.5057	4.954
.5540	.5530	3.481	.9981	16.23	16.26	.9944	.06150	6.962	527.7	527.7	.5066	.5056	4.954
.5550	.5540	3.487	.9981	16.33	16.36	.9945	.06112	6.974	534.4	534.4	.5065	.5056	4.953
.5560	.5550	3.493	.9982	16.44	16.47	.9945	.06074	6.987	541.2	541.2	.5065	.5055	4.953
.5570	.5560	3.500	.9982	16.54	16.57	.9946	.06036	6.999	548.0	548.0	.5064	.5055	4.953
.5580	.5570	3.506	.9982	16.64	16.67	.9946	.05998	7.012	554.9	554.9	.5063	.5054	4.953
.5590	.5580	3.512	.9982	16.75	16.78	.9947	.05960	7.025	561.9	561.9	.5063	.5054	4.953
.5600	.5590	3.519	.9982	16.85	16.88	.9948	.05923	7.037	569.1	569.1	.5062	.5053	4.952
.5610	.5600	3.525	.9983	16.96	16.99	.9948	.05886	7.050	576.3	576.3	.5061	.5052	4.952
.5620	.5610	3.531	.9983	17.07	17.10	.9949	.05849	7.062	583.5	583.5	.5061	.5052	4.952
.5630	.5621	3.537	.9983	17.17	17.20	.9949	.05813	7.075	590.9	590.9	.5060	.5051	4.952
.5640	.5631	3.544	.9983	17.28	17.31	.9950	.05776	7.087	598.4	598.4	.5059	.5051	4.951
.5650	.5641	3.550	.9984	17.39	17.42	.9950	.05740	7.100	606.0	606.0	.5059	.5050	4.951
.5660	.5651	3.556	.9984	17.50	17.53	.9951	.05704	7.113	613.6	613.6	.5058	.5050	4.951
.5670	.5661	3.563	.9984	17.61	17.64	.9951	.05669	7.125	621.4	621.4	.5057	.5049	4.951
.5680	.5671	3.569	.9984	17.72	17.75	.9952	.05633	7.138	629.2	629.2	.5057	.5049	4.951
.5690	.5681	3.575	.9984	17.84	17.86	.9952	.05598	7.150	637.3	637.3	.5056	.5048	4.950
.5700	.5691	3.581	.9985	17.95	17.98	.9953	.05563	7.163	645.2	645.2	.5056	.5048	4.950
.5710	.5701	3.588	.9985	18.06	18.09	.9953	.05528	7.175	653.4	653.4	.5055	.5047	4.950
.5720	.5711	3.594	.9985	18.18	18.20	.9954	.05494	7.188	661.7	661.7	.5054	.5047	4.950
.5730	.5722	3.600	.9985	18.29	18.32	.9954	.05459	7.201	670.0	670.0	.5054	.5046	4.950
.5740	.5732	3.607	.9985	18.41	18.43	.9955	.05425	7.213	678.5	678.5	.5053	.5046	4.949
.5750	.5742	3.613	.9986	18.52	18.55	.9955	.05391	7.226	687.1	687.1	.5053	.5045	4.949
.5760	.5752	3.619	.9986	18.64	18.67	.9956	.05358	7.238	695.8	695.8	.5052	.5045	4.949
.5770	.5762	3.625	.9986	18.76	18.78	.9956	.05324	7.251	704.6	704.6	.5051	.5044	4.949
.5780	.5772	3.632	.9986	18.88	18.90	.9957	.05291	7.263	713.5	713.5	.5051	.5044	4.949
.5790	.5782	3.638	.9986	18.99	19.02	.9957	.05258	7.276	722.5	722.5	.5050	.5043	4.949
.5800	.5792	3.644	.9986	19.11	19.14	.9957	.05225	7.289	731.6	731.6	.5050	.5043	4.948
.5810	.5802	3.651	.9987	19.23	19.26	.9958	.05192	7.301	740.9	740.9	.5049	.5043	4.948
.5820	.5812	3.657	.9987	19.36	19.38	.9958	.05160	7.314	750.3	750.3	.5049	.5042	4.948
.5830	.5822	3.663	.9987	19.48	19.50	.9959	.05127	7.326	759.8	759.8	.5048	.5042	4.948
.5840	.5832	3.669	.9987	19.60	19.63	.9959	.05095	7.339	769.4	769.4	.5048	.5041	4.948
.5850	.5843	3.676	.9987	19.73	19.75	.9960	.05063	7.351	779.1	779.1	.5047	.5041	4.948
.5860	.5853	3.682	.9987	19.85	19.87	.9960	.05032	7.364	788.9	788.9	.5047	.5040	4.947
.5870	.5863	3.688	.9988	19.97	20.00	.9960	.05000	7.376	798.9	798.9	.5046	.5040	4.947
.5880	.5873	3.695	.9988	20.10	20.13	.9961	.04969	7.389	809.0	809.0	.5046	.5039	4.947
.5890	.5883	3.701	.9988	20.23	20.25	.9961	.04938	7.402	819.3	819.3	.5045	.5039	4.947

Table C–2 – Continued

d/L	d/Lo	$\frac{2\pi d}{L}$	tanh $\frac{2\pi d}{L}$	sinh $\frac{2\pi d}{L}$	cosh $\frac{2\pi d}{L}$	H/Ho'	K	$\frac{4\pi d}{L}$	sinh $\frac{4\pi d}{L}$	cosh $\frac{4\pi d}{L}$	n	C_G/C_o	M
.5900	.5893	3.707	.9988	20.36	20.38	.9962	.04907	7.414	829.6	829.6	.5045	.5039	4.947
.5910	.5903	3.713	.9988	20.48	20.51	.9962	.04876	7.427	840.1	840.1	.5044	.5038	4.947
.5920	.5913	3.720	.9988	20.61	20.64	.9962	.04846	7.439	850.7	850.7	.5044	.5038	4.946
.5930	.5923	3.726	.9988	20.74	20.77	.9963	.04815	7.452	861.5	861.5	.5043	.5037	4.946
.5940	.5933	3.732	.9989	20.87	20.90	.9963	.04785	7.464	872.4	872.4	.5043	.5037	4.946
.5950	.5943	3.739	.9989	21.01	21.03	.9964	.04755	7.477	883.4	883.4	.5042	.5037	4.946
.5960	.5953	3.745	.9989	21.14	21.16	.9964	.04725	7.490	894.6	894.6	.5042	.5036	4.946
.5970	.5963	3.751	.9989	21.27	21.30	.9964	.04696	7.502	905.9	905.9	.5041	.5036	4.946
.5980	.5974	3.757	.9989	21.41	21.43	.9965	.04667	7.515	917.3	917.3	.5041	.5036	4.946
.5990	.5984	3.764	.9989	21.54	21.55	.9965	.04639	7.527	929.0	929.0	.5041	.5035	4.945
.6000	.5994	3.770	.9989	21.68	21.70	.9966	.04609	7.540	940.7	940.7	.5040	.5035	4.945
.6100	.6094	3.833	.9991	23.08	23.11	.9970	.04328	7.666	1067.	1067.	.5036	.5031	4.944
.6200	.6195	3.896	.9992	24.58	24.60	.9972	.04065	7.791	1210.	1210.	.5032	.5028	4.943
.6300	.6295	3.958	.9993	26.18	26.20	.9975	.03817	7.917	1371.	1371.	.5029	.5025	4.942
.6400	.6396	4.021	.9994	27.88	27.89	.9978	.03585	8.043	1555.	1555.	.5026	.5023	4.941
.6500	.6496	4.084	.9994	29.69	29.70	.9980	.03367	8.163	1754.	1754.	.5023	.5020	4.941
.6600	.6597	4.147	.9995	31.61	31.63	.9982	.03162	8.294	1999.	1999.	.5021	.5018	4.940
.6700	.6697	4.210	.9996	33.66	33.68	.9984	.02969	8.419	2267.	2267.	.5019	.5016	4.939
.6800	.6797	4.273	.9996	35.85	35.86	.9985	.02789	8.545	2571.	2571.	.5017	.5015	4.939
.6900	.6898	4.335	.9997	38.17	38.18	.9987	.02619	8.671	2915.	2915.	.5015	.5013	4.938
.7000	.6998	4.398	.9997	40.65	40.66	.9988	.02459	8.796	3305.	3305.	.5013	.5012	4.938
.7100	.7098	4.461	.9997	43.29	43.30	.9989	.02310	8.922	3748.	3748.	.5012	.5011	4.938
.7200	.7198	4.524	.9998	46.09	46.10	.9990	.02169	9.048	4250.	4250.	.5011	.5010	4.937
.7300	.7299	4.587	.9998	49.08	49.09	.9991	.02037	9.173	4819.	4819.	.5010	.5009	4.937
.7400	.7399	4.650	.9998	52.27	52.28	.9992	.01913	9.299	5464.	5464.	.5009	.5008	4.937
.7500	.7499	4.712	.9998	55.66	55.66	.9993	.01796	9.425	6195.	6195.	.5008	.5007	4.936
.7600	.7599	4.775	.9999	59.26	59.27	.9994	.01687	9.550	7025.	7025.	.5007	.5006	4.936
.7700	.7699	4.838	.9999	63.11	63.12	.9995	.01584	9.676	7966.	7966.	.5006	.5005	4.936
.7800	.7799	4.901	.9999	67.20	67.21	.9995	.01488	9.802	9032.	9032.	.5005	.5005	4.936
.7900	.7899	4.964	.9999	71.56	71.56	.9996	.01397	9.927	10240.	10240.	.5005	.5004	4.936
.8000	.7999	5.027	.9999	76.21	76.21	.9996	.01312	10.05	11610.	11610.	.5004	.5004	4.936
.8100	.8099	5.089	.9999	81.14	81.14	.9997	.01232	10.18	13170.	13170.	.5004	.5004	4.936
.8200	.8199	5.152	.9999	86.40	86.40	.9997	.01157	10.30	14930.	14930.	.5003	.5003	4.936
.8300	.8300	5.215	.9999	92.01	92.01	.9997	.01087	10.43	16930.	16930.	.5003	.5003	4.935
.8400	.8400	5.278	1.000	97.98	97.98	.9998	.01021	10.56	19200.	19200.	.5003	.5003	4.935
.8500	.8500	5.341	1.000	104.3	104.3	.9998	.009585	10.68	21770.	21770.	.5002	.5002	4.935
.8600	.8600	5.404	1.000	111.1	111.1	.9998	.009000	10.81	24680.	24680.	.5002	.5002	4.935
.8700	.8700	5.466	1.000	118.3	118.3	.9998	.008453	10.93	27990.	27990.	.5002	.5002	4.935
.8800	.8800	5.529	1.000	126.0	126.0	.9998	.007939	11.06	31730.	31730.	.5002	.5002	4.935
.8900	.8900	5.592	1.000	134.1	134.1	.9999	.007455	11.18	35980.	35980.	.5002	.5002	4.935
.9000	.9000	5.655	1.000	142.8	142.8	.9999	.007001	11.31	40800.	40800.	.5001	.5001	4.935
.9100	.9100	5.718	1.000	152.1	152.1	.9999	.006575	11.44	46260.	46260.	.5001	.5001	4.935
.9200	.9200	5.781	1.000	162.0	162.0	.9999	.006174	11.56	52460.	52460.	.5001	.5001	4.935
.9300	.9300	5.843	1.000	172.5	172.5	.9999	.005798	11.69	59480.	59480.	.5001	.5001	4.935
.9400	.9400	5.906	1.000	183.7	183.7	.9999	.005445	11.81	67450.	67450.	.5001	.5001	4.935
.9500	.9500	5.969	1.000	195.6	195.6	.9999	.005114	11.94	76480.	76480.	.5001	.5001	4.935
.9600	.9600	6.032	1.000	208.2	208.2	.9999	.004802	12.06	86720.	86720.	.5001	.5001	4.935
.9700	.9700	6.095	1.000	221.7	221.7	.9999	.004510	12.19	98340.	98340.	.5001	.5001	4.935
.9800	.9800	6.158	1.000	236.1	236.1	.9999	.004235	12.32	111500.	111500.	.5001	.5001	4.935
.9900	.9900	6.220	1.000	251.4	251.4	1.0000	.003977	12.44	126400.	126400.	.5000	.5000	4.935
1.000	1.000	6.283	1.000	267.7	267.7	1.0000	.003735	12.57	143400.	143400.	.5000	.5000	4.935

after Wiegel, R. L., "Oscillatory Waves," U.S. Army, Beach Erosion Board, Bulletin, Special Issue No. 1, July 1948.

Table C-3. Deepwater Wave Length (L_o) and Velocity (C_o) as A Function of Wave Period.

T (Seconds)	C_o (Ft./Sec)	C_o (Knots)	L_o (Feet)	T (Seconds)	C_o (Ft./Sec)	C_o (Knots)	L_o (Feet)	T (Seconds)	C_o (Ft./Sec)	C_o (Knots)	L_o (Feet)	T (Seconds)	C_o (Ft./Sec)	C_o (Knots)	L_o (Feet)
3.0	15.4	9.1	46.1	7.6	38.9	23.0	296	12.2	62.4	37.0	762	16.8	86.0	50.9	1444
3.1	15.9	9.4	49.2	7.7	39.4	23.3	304	12.3	63.0	37.3	775	16.9	86.5	51.2	1461
3.2	16.4	9.7	52.4	7.8	39.9	23.6	312	12.4	63.5	37.6	787	17.0	87.0	51.5	1479
3.3	16.9	10.0	55.8	7.9	40.4	23.9	320	12.5	64.0	37.9	800	17.1	87.5	51.8	1496
3.4	17.4	10.3	59.2	8.0	40.9	24.2	328	12.6	64.5	38.2	813	17.2	88.0	52.1	1514
3.5	17.9	10.6	62.7	8.1	41.4	24.5	336	12.7	65.0	38.5	826	17.3	88.5	52.4	1531
3.6	18.4	10.9	66.4	8.2	42.0	24.8	344	12.8	65.5	38.8	839	17.4	89.0	52.7	1549
3.7	18.9	11.2	70.1	8.3	42.5	25.1	353	12.9	66.0	39.1	852	17.5	89.6	53.0	1567
3.8	19.4	11.5	73.9	8.4	43.0	25.4	361	13.0	66.5	39.4	865	17.6	90.1	53.3	1585
3.9	20.0	11.8	77.9	8.5	43.5	25.7	370	13.1	67.0	39.7	879	17.7	90.6	53.6	1603
4.0	20.5	12.1	81.9	8.6	44.0	26.1	379	13.2	67.6	40.0	892	17.8	91.1	53.9	1621
4.1	21.0	12.4	86.1	8.7	44.5	26.4	388	13.3	68.1	40.3	906	17.9	91.6	54.2	1639
4.2	21.5	12.7	90.3	8.8	45.0	26.7	397	13.4	68.6	40.6	919	18.0	92.1	54.5	1658
4.3	22.0	13.0	94.7	8.9	45.6	27.0	406	13.5	69.1	40.9	933	18.1	92.6	54.8	1677
4.4	22.5	13.3	99.1	9.0	46.1	27.3	415	13.6	69.6	41.2	947	18.2	93.1	55.1	1695
4.5	23.0	13.6	104	9.1	46.6	27.6	424	13.7	70.1	41.5	961	18.3	93.6	55.4	1714
4.6	23.5	13.9	108	9.2	47.1	27.9	433	13.8	70.6	41.8	975	18.4	94.2	55.8	1732
4.7	24.0	14.2	113	9.3	47.6	28.2	442	13.9	71.1	42.1	989	18.5	94.7	56.1	1751
4.8	24.6	14.5	118	9.4	48.1	28.5	452	14.0	71.6	42.4	1004	18.6	95.2	56.4	1770
4.9	25.1	14.8	123	9.5	48.6	28.8	462	14.1	72.2	42.7	1018	18.7	95.7	56.7	1789
5.0	25.6	15.2	128	9.6	49.1	29.1	472	14.2	72.7	43.0	1032	18.8	96.2	57.0	1809
5.1	26.1	15.5	133	9.7	49.6	29.4	482	14.3	73.2	43.3	1047	18.9	96.7	57.3	1828
5.2	26.6	15.8	138	9.8	50.2	29.7	492	14.4	73.7	43.6	1062	19.0	97.2	57.6	1847
5.3	27.1	16.1	144	9.9	50.7	30.0	502	14.5	74.2	43.9	1076	19.1	97.8	57.9	1867
5.4	27.6	16.4	149	10.0	51.2	30.3	512	14.6	74.7	44.2	1091	19.2	98.3	58.2	1886
5.5	28.1	16.7	155	10.1	51.7	30.6	522	14.7	75.2	44.5	1106	19.3	98.8	58.5	1906
5.6	28.7	17.0	161	10.2	52.2	30.9	533	14.8	75.7	44.8	1121	19.4	99.3	58.8	1926
5.7	29.2	17.3	166	10.3	52.7	31.2	543	14.9	76.2	45.1	1137	19.5	99.8	59.1	1946
5.8	29.7	17.6	172	10.4	53.2	31.5	554	15.0	76.8	45.5	1152	19.6	100.3	59.4	1966
5.9	30.2	17.9	178	10.5	53.7	31.8	564	15.1	77.3	45.8	1167	19.7	100.8	59.7	1986
6.0	30.7	18.2	184	10.6	54.2	32.1	575	15.2	77.8	46.1	1183	19.8	101.3	60.0	2006
6.1	31.2	18.5	191	10.7	54.8	32.4	586	15.3	78.3	46.4	1199	19.9	101.8	60.3	2027
6.2	31.7	18.8	197	10.8	55.3	32.7	597	15.4	78.8	46.7	1214	20.0	102.4	60.6	2047
6.3	32.2	19.1	203	10.9	55.8	33.0	608	15.5	79.3	47.0	1230	21.0	107.5	63.6	2257
6.4	32.8	19.4	210	11.0	56.3	33.3	620	15.6	79.8	47.3	1246	22.0	112.6	66.7	2477
6.5	33.3	19.7	216	11.1	56.8	33.6	631	15.7	80.4	47.6	1263	23.0	117.7	69.7	2707
6.6	33.8	20.0	223	11.2	57.3	33.9	642	15.8	80.9	47.9	1277	24.0	122.8	72.7	2948
6.7	34.3	20.3	230	11.3	57.8	34.2	654	15.9	81.4	48.2	1293	25.0	128.0	75.7	3199
6.8	34.8	20.6	237	11.4	58.3	34.5	665	16.0	81.9	48.5	1310	26.0	133.1	78.8	3460
6.9	35.3	20.9	244	11.5	58.9	34.8	677	16.1	82.4	48.8	1326				
7.0	35.8	21.2	251	11.6	59.4	35.1	689	16.2	82.9	49.1	1343				
7.1	36.3	21.5	258	11.7	59.9	35.4	701	16.3	83.4	49.4	1359				
7.2	36.8	21.8	265	11.8	60.4	35.8	713	16.4	83.9	49.7	1376				
7.3	37.4	22.1	273	11.9	60.9	36.1	725	16.5	84.4	50.0	1393				
7.4	37.9	22.4	280	12.0	61.4	36.4	737	16.6	85.0	50.3	1410				
7.5	38.4	22.7	288	12.1	61.9	36.7	750	16.7	85.5	50.6	1427				

$$L_o = \frac{gT^2}{2\pi} = 5.12\,T^2$$

$$C_o = \frac{gT}{2\pi} = 5.12\,T$$

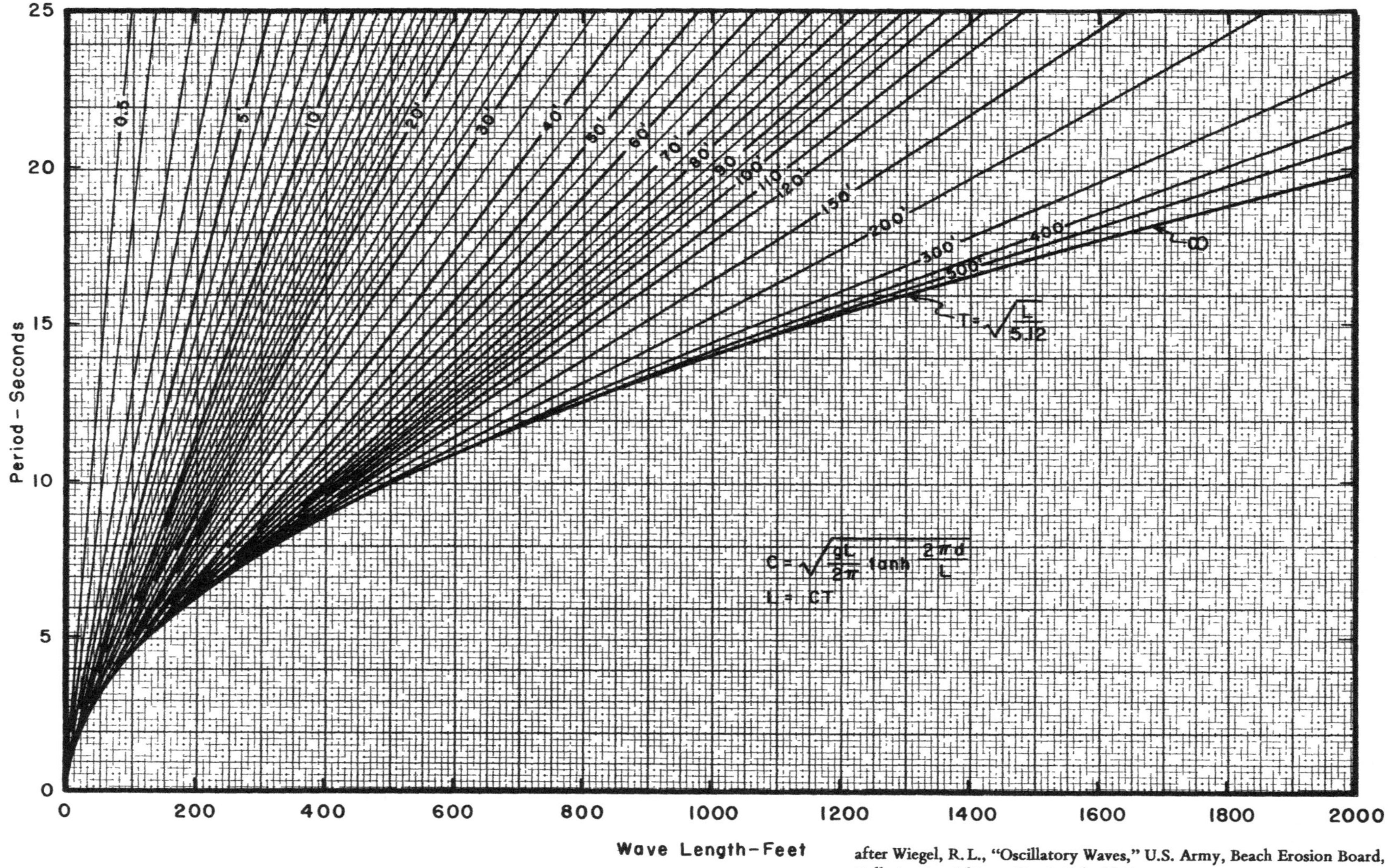

Plate C-2. Relationship Between Wave Period, Length and Depth

after Wiegel, R. L., "Oscillatory Waves," U.S. Army, Beach Erosion Board, Bulletin, Special Issue No. 1, July 1948.

$$c = \sqrt{\frac{gL}{2\pi} \tanh \frac{2\pi d}{L}}$$

$$L = cT$$

$$T = \sqrt{\frac{L}{5.12}}$$

Wave Length—Feet

Period — Seconds

The axis labels: Period Seconds 0, 5, 10, 15, 20, 25. Wave Length Feet 0, 200, 400, 600, 800, 1000, 1200, 1400, 1600, 1800, 2000.

C-3

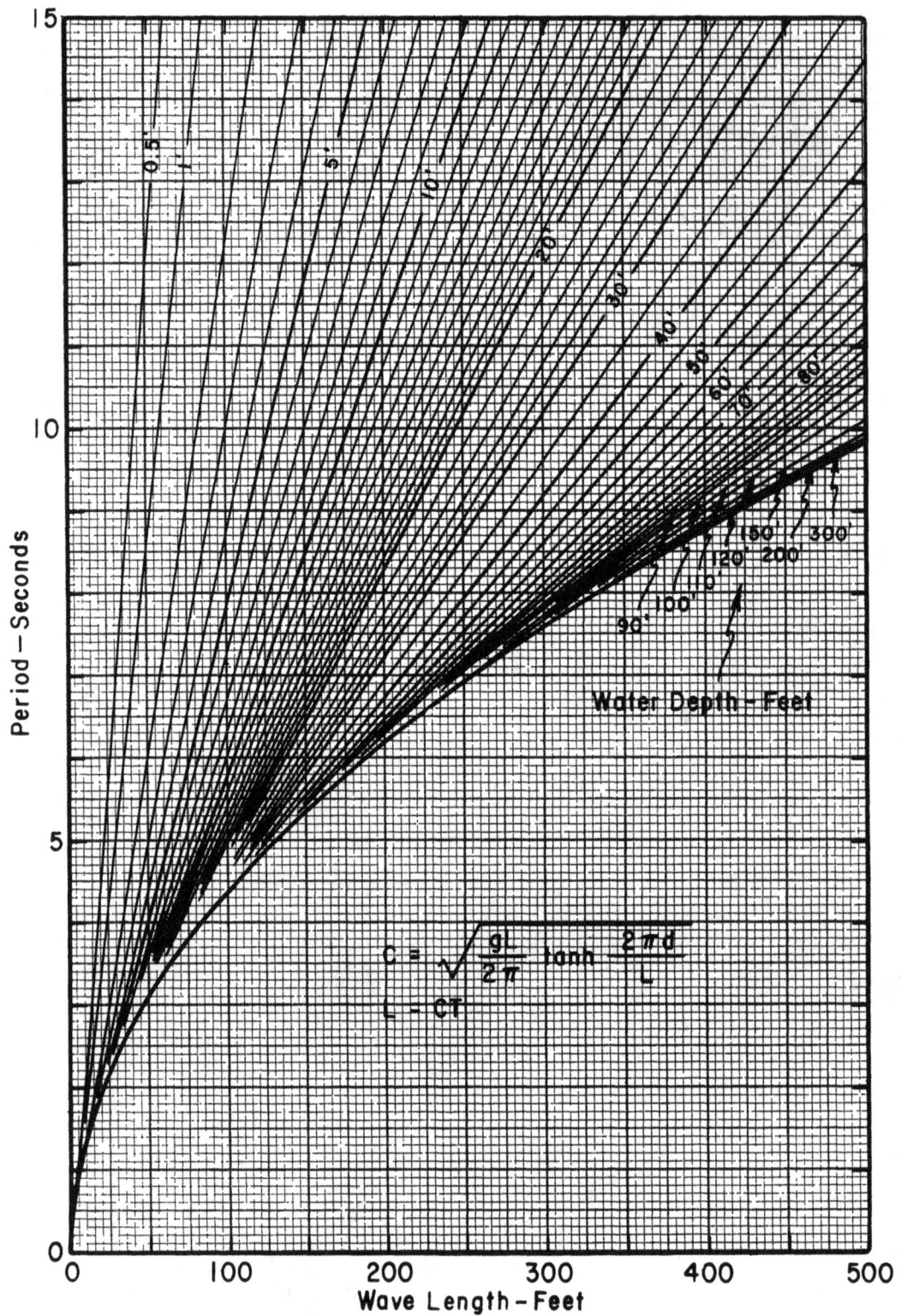

Plate C-3. Relationship Between Wave Period, Length, and Depth

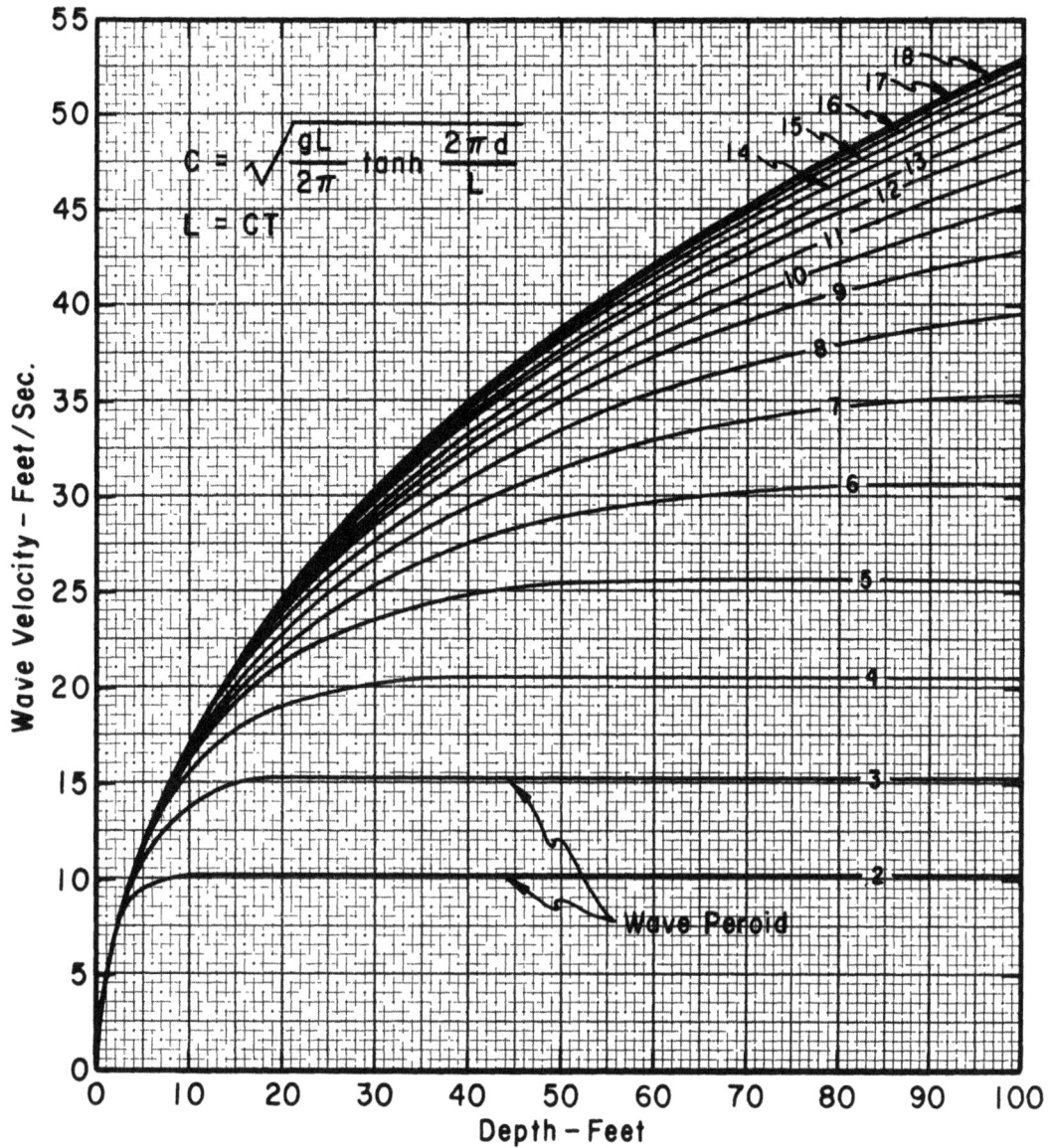

after Wiegel, R.L., "Oscillatory Waves," U.S. Army, Beach Erosion Board, Bulletin, Special Issue No. 1, July 1948.

Plate C-4. Relationship Between Wave Period, Velocity, and Depth

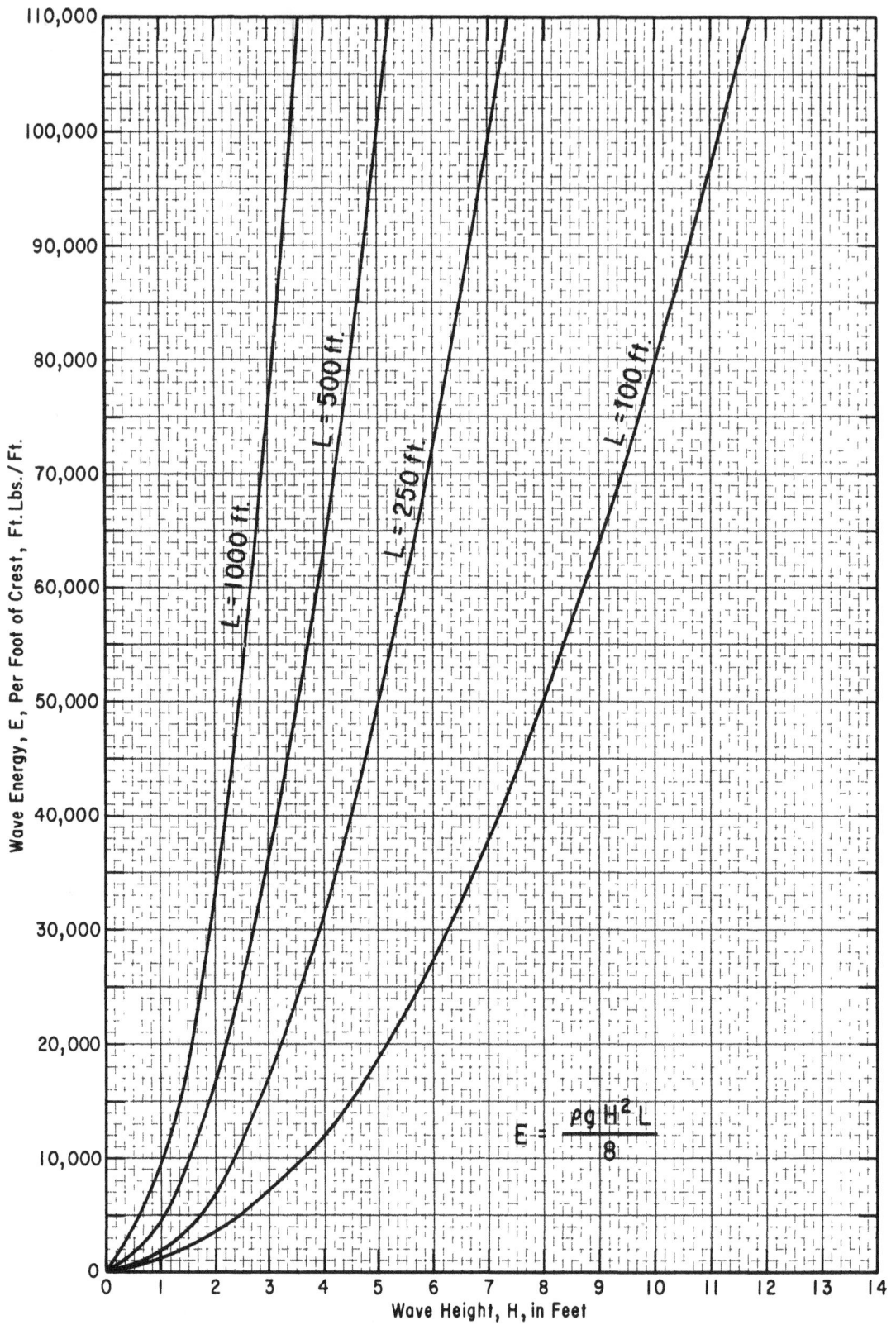

Wave Energy, E, Per Foot of Crest, Ft. Lbs./ Ft.

Wave Height, H, in Feet

L = 1000 ft.

L = 500 ft.

L = 250 ft.

L = 100 ft.

$$E = \frac{\rho g H^2 L}{8}$$

Plate C-5. Relationship Between Wave Energy, Wave Length and Wave Height

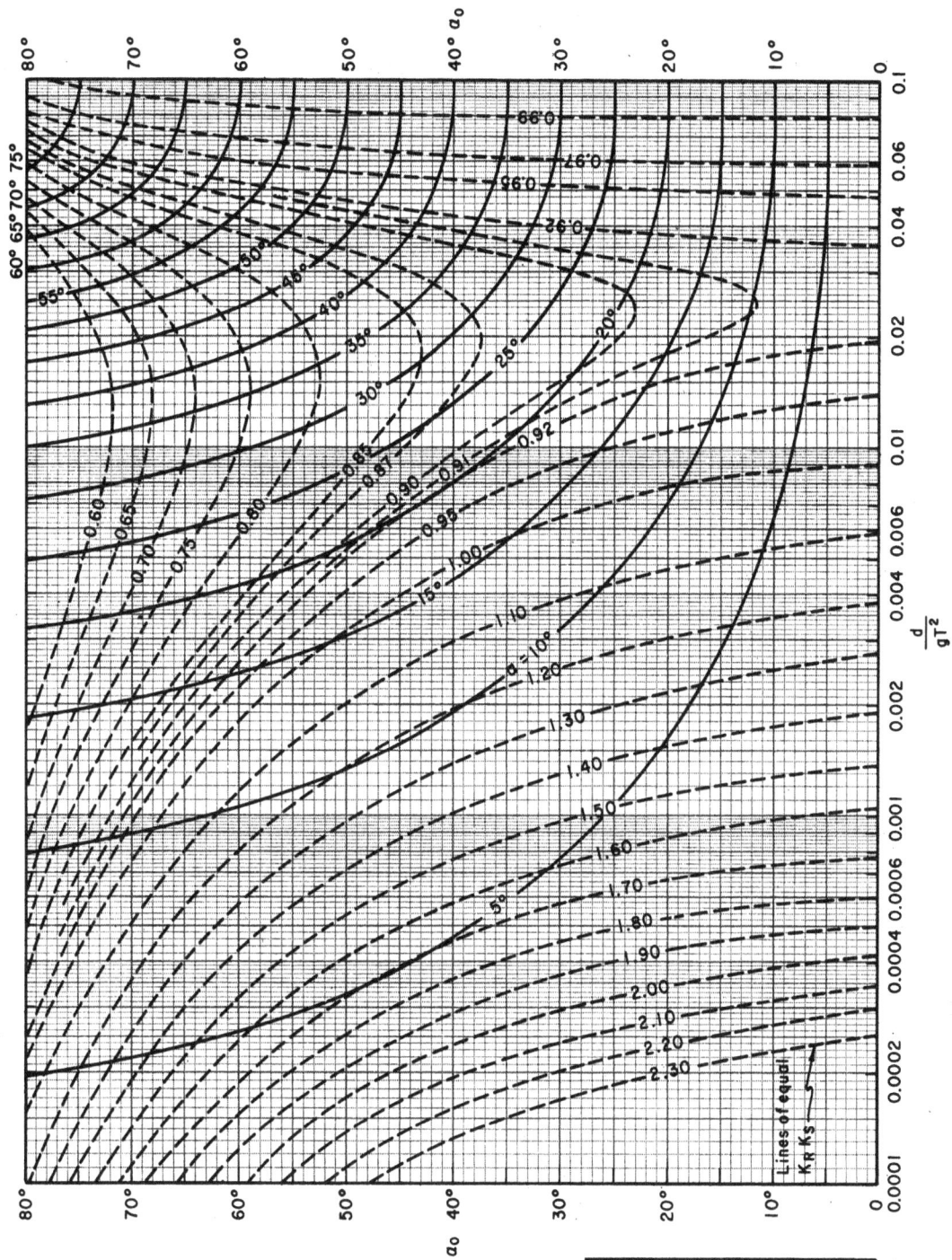

Plate C-6. Change in Wave Direction and Height Due to Refraction on Slopes with Straight, Parallel Depth Contours Including Shoaling

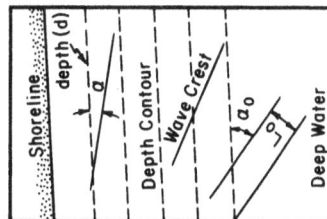

Table C—4. Values Used for Plotting Orthogonals.

Wave period (T) = 5 seconds, Deep water wave length = (L_o) = 128 feet

Depth (Fathoms)	$\Delta L/L_{av}$	C_d/C_s*	C_s/C_d*
10	0.005	1.004	0.996
9	0.007	1.007	0.993
8	0.011	1.012	0.988
7	0.019	1.020	0.980
6	0.032	1.034	0.968
5	0.052	1.055	0.948
4	0.085	1.090	0.917
3	0.145	1.158	0.864
2	0.290	1.341	0.746
1			

Wave period (T) = 6 seconds, Deep water wave length = (L_o) = 184.2 ft.

Depth (Fathoms)	$\Delta L/L_{av}$	C_d/C_s*	C_s/C_d*
15	0.005	1.005	0.995
13	0.011	1.011	0.989
11	0.009	1.009	0.991
10	0.014	1.013	0.987
9	0.019	1.019	0.981
8	0.026	1.026	0.974
7	0.037	1.037	0.964
6	0.050	1.052	0.950
5	0.074	1.075	0.930
4	0.095	1.111	0.900
3	0.154	1.180	0.847
2	0.330	1.363	0.734
1			

Wave period (T) = 7 seconds, Deep water wave length = (L_o) = 251 feet

Depth (Fathoms)	$\Delta L/L_{av}$	C_d/C_s*	C_s/C_d*
21	0.006	1.006	0.994
17.5	0.010	1.010	0.990
15	0.014	1.015	0.985
13	0.025	1.025	0.976
11	0.018	1.018	0.982
10	0.023	1.023	0.977
9	0.029	1.030	0.971
8	0.037	1.039	0.964
7	0.047	1.049	0.953
6	0.064	1.065	0.939
5	0.082	1.086	0.921
4	0.117	1.123	0.890
3	0.175	1.192	0.839
2	0.310	1.378	0.726
1			

Wave period (T) = 8 seconds, Deep water wave length (L_o) = 327 feet

Depth (Fathoms)	$\Delta L/L_{av}$	C_d/C_s*	C_s/C_d*
27	0.006	1.015	0.986
20	0.013	1.013	0.987
17.5	0.021	1.021	0.979
15	0.013	1.012	0.988
14	0.014	1.014	0.986
13	0.016	1.017	0.983
12	0.020	1.021	0.980
11	0.025	1.025	0.975
10	0.031	1.031	0.970
9	0.038	1.037	0.964
8	0.046	1.046	0.956
7	0.056	1.059	0.946
6	0.069	1.069	0.933
5	0.090	1.095	0.914
4	0.124	1.131	0.844
3	0.183	1.199	0.834
2	0.30	1.389	0.720
1			

* Note:- C_d is the wave velocity in deeper water; C_s is the wave velocity in shallower water. When an orthogonal is being drawn from deep to shallower water, $C_d/C_s = C_1/C_2$. When an orthogonal is being drawn from shallow to deep water, $C_s/C_d = C_1/C_2$.

Table C—4 — Continued

Wave period (T) = 9 seconds — Deep water wave length (L_o) = 415 feet

Depth (fathoms)	$\Delta L/L_{av}$	c_d/c_s*	c_s/c_d*
55	0.011	1.005	0.995
30	0.012	1.011	0.989
25	0.025	1.026	0.975
20	0.022	1.022	0.978
17.5	0.053	1.032	0.969
15	0.015	1.017	0.983
14	0.019	1.019	0.981
13	0.022	1.022	0.978
12	0.025	1.026	0.975
11	0.029	1.031	0.970
10	0.035	1.036	0.966
9	0.043	1.042	0.959
8	0.048	1.051	0.952
7	0.064	1.062	0.941
6	0.071	1.077	0.929
5	0.098	1.100	0.909
4	0.136	1.136	0.881
3	0.196	1.206	0.829
2	0.350	1.390	0.720
1			

Wave period (T) = 10 seconds — Deep water wave length (L_o) = 512 feet

Depth (fathoms)	$\Delta L/L_{av}$	c_d/c_s*	c_s/c_d*
43	0.010	1.007	0.993
35	0.010	1.011	0.989
30	0.020	1.021	0.979
25	0.040	1.040	0.962
20	0.029	1.031	0.970
17.5	0.041	1.042	0.960
15	0.021	1.020	0.980
14	0.023	1.023	0.978
13	0.026	1.026	0.975
12	0.029	1.030	0.971
11	0.034	1.034	0.967
10	0.039	1.040	0.962
9	0.045	1.046	0.956
8	0.054	1.055	0.948
7	0.065	1.066	0.938
6	0.078	1.081	0.925
5	0.100	1.103	0.906
4	0.132	1.140	0.878
3	0.189	1.211	0.826
2	0.326	1.397	0.716
1			

Wave period (T) = 11 seconds — Deep water wave length (L_o) = 620 feet

Depth (fathoms)	$\Delta L/L_{av}$	c_d/c_s*	c_s/c_d*
50	0.01	1.010	0.990
40	0.01	1.012	0.989
35	0.02	1.019	0.982
30	0.03	1.032	0.969
25	0.05	1.053	0.950
20	0.04	1.037	0.964
17.5	0.05	1.049	0.953
15	0.020	1.023	0.978
14	0.027	1.026	0.975
13	0.029	1.029	0.972
12	0.032	1.033	0.968
11	0.038	1.037	0.964
10	0.044	1.043	0.959
9	0.050	1.049	0.954
8	0.058	1.058	0.947
7	0.070	1.068	0.936
6	0.025	1.084	0.923
5	0.105	1.105	0.905
4	0.137	1.143	0.875
3	0.203	1.208	0.825
2	0.325	1.400	0.714
1			

Wave period (T) = 12 seconds — Deep water wave length (L_o) = 737 feet

Depth (fathoms)	$\Delta L/L_{av}$	c_d/c_s*	c_s/c_d*
61	0.007	1.007	0.994
50	0.010	1.007	0.993
45	0.008	1.011	0.989
40	0.012	1.018	0.983
35	0.032	1.027	0.974
30	0.040	1.042	0.960
25	0.065	1.062	0.941
20	0.044	1.043	0.959
17.5	0.051	1.054	0.949
15	0.024	1.025	0.976
14	0.027	1.028	0.973
13	0.031	1.031	0.970
12	0.035	1.034	0.967
11	0.039	1.039	0.962
10	0.044	1.044	0.958
9	0.049	1.051	0.951
8	0.057	1.059	0.944
7	0.068	1.070	0.935
6	0.083	1.086	0.921
5	0.105	1.108	0.904
4	0.135	1.148	0.871
3	0.20	1.210	0.829
2	0.33	1.406	0.709
1			

Table C-4 – Continued

Wave period (T) = 13 seconds
Deep water wave length (L_o) = 865 feet

Depth (fathoms)	$\Delta L/L_{av}$	c_d/c_s*	c_s/c_d*	Depth (fathoms)	$\Delta L/L_{av}$	c_d/c_s*	c_s/c_d*
72	0.006	1.006	0.994	12	0.036	1.036	0.966
60	0.013	1.013	0.987	11	0.040	1.041	0.961
50	0.012	1.011	0.989	10	0.045	1.046	0.956
45	0.016	1.017	0.984	9	0.051	1.052	0.950
40	0.024	1.024	0.977	8	0.059	1.061	0.942
35	0.033	1.034	0.967	7	0.070	1.071	0.933
30	0.048	1.049	0.953	6	0.085	1.088	0.919
25	0.070	1.072	0.933	5	0.103	1.109	0.902
20	0.047	1.047	0.955	4	0.135	1.147	0.872
17.5	0.055	1.058	0.945	3	0.200	1.213	0.824
15	0.027	1.027	0.974	2	0.341	1.410	0.709
14	0.029	1.029	0.972	1			
13	0.032	1.033	0.968				
12							

Wave period (T) = 14 seconds
Deep water wave length (L_o) = 1,003.5 feet

Depth (fathoms)	$\Delta L/L_{av}$	c_d/c_s*	c_s/c_d*	Depth (fathoms)	$\Delta L/L_{av}$	c_d/c_s*	c_s/c_d*
85	0.006	1.007	0.993	13	0.033	1.033	0.967
70	0.030	1.030	0.970	12	0.037	1.037	0.964
50	0.038	1.038	0.963	11	0.041	1.042	0.960
40	0.029	1.030	0.971	10	0.046	1.047	0.955
35	0.040	1.041	0.961	9	0.052	1.054	0.949
30	0.055	1.056	0.947	8	0.058	1.062	0.942
25	0.075	1.078	0.928	7	0.070	1.073	0.932
20	0.049	1.050	0.952	6	0.085	1.089	0.918
17.5	0.059	1.061	0.942	5	0.105	1.139	0.902
15	0.028	1.029	0.972	4	0.144	1.148	0.871
14	0.030	1.031	0.970	3	0.188	1.217	0.822
13				2	0.338	1.405	0.712
				1			

Table C-4 – Continued

Wave period (T) = 15 seconds
Deep water wave length (L_o) = 1,152 feet

Depth (Fathoms)	L/L_{av}	c_d/c_s*	c_s/c_d*	Depth (Fathoms)	L/L_{av}	c_d/c_s*	c_s/c_d*
96	0.060	1.060	0.944	12	0.058	1.038	0.963
50	0.047	1.048	0.954	11	0.043	1.043	0.969
40	0.080	1.083	0.923	10	0.047	1.048	0.954
30	0.059	1.061	0.942	9	0.053	1.054	0.948
25	0.080	1.084	0.925	8	0.061	1.063	0.941
20	0.050	1.053	0.949	7	0.073	1.076	0.930
17.5	0.063	1.064	0.940	6	0.080	1.089	0.918
15	0.029	1.029	0.972	5	0.106	1.112	0.899
14	0.031	1.032	0.969	4	0.138	1.148	0.871
13	0.054	1.035	0.966	3	0.197	1.218	0.821
12				2	0.356	1.405	0.712
				1			

Wave period (T) = 16 seconds
Deep water wave length (L_o) = 1,310 feet

Depth (Fathoms)	$\Delta L/L_{av}$	c_d/c_s*	c_s/c_d*	Depth (Fathoms)	$\Delta L/L_{av}$	c_d/c_s*	c_s/c_d*
109	0.028	1.028	0.973	12	0.038	1.039	0.962
70	0.055	1.057	0.946	11	0.042	1.044	0.958
50	0.055	1.056	0.947	10	0.046	1.049	0.953
40	0.038	1.039	0.962	9	0.053	1.056	0.947
35	0.049	1.050	0.952	8	0.063	1.063	0.941
30	0.065	1.066	0.938	7	0.074	1.075	0.930
25	0.084	1.088	0.919	6	0.088	1.090	0.917
20	0.051	1.055	0.948	5	0.108	1.114	0.898
17.5	0.067	1.066	0.938	4	0.133	1.149	0.870
15	0.030	1.030	0.971	3	0.200	1.215	0.823
14	0.033	1.032	0.969	2	0.344	1.407	0.711
13	0.035	1.035	0.966	1			
12	0.055						

C-39

Table C-4 – Continued

Wave period (T) = 17 seconds
Deep water wave length (L_o) = 1,480 feet

Depth (fathoms)	L/L_{av}	c_d/c_s*	c_s/c_d*
123	0.007	1.008	0.992
100	0.036	1.057	0.965
70	0.067	1.070	0.935
50	0.060	1.063	0.941
40	0.044	1.043	0.959
35	0.052	1.054	0.949
30	0.065	1.069	0.935
25	0.091	1.092	0.916
20	0.055	1.057	0.946
17.5	0.065	1.068	0.937
15	0.077	1.083	0.923
12.5	0.102	1.106	0.904
10	0.135	1.142	0.876
7.5	0.195	1.211	0.826
5	0.112	1.115	0.897
4	0.140	1.147	0.872
3	0.189	1.222	0.819
2	0.339	1.400	0.714
1			

Wave period (T) = 18 seconds
Deep water wave length (L_o) = 1,659 ft.

Depth (fathoms)	L/L_{av}	c_d/c_s*	c_s/c_d*
138	0.015	1.016	0.984
100	0.047	1.048	0.954
70	0.079	1.081	0.925
50	0.067	1.069	0.935
40	0.099	1.106	0.904
30	0.069	1.073	0.932
25	0.095	1.095	0.913
20	0.125	1.132	0.884
15	0.181	1.199	0.834
10	0.106	1.111	0.900
8	0.126	1.145	0.873
6	0.117	1.091	0.917
5	0.10	1.112	0.899
4	0.14	1.155	0.866
3	0.20	1.219	0.820
2	0.73	1.409	0.710
1			

Table C–5. Conversion Factors — British to Metric Units of Measurement

The following conversion factors adopted by the Bureau of Reclamation are those published by the American Society for Testing and Materials (ASTM Metric Practice Guide, January 1964) except that additional factors (*) commonly used in the Bureau have been added. Further discussion of definitions of quantities and units is given on pages 10-11 of the ASTM Metric Practice Guide.

The metric units and conversion factors adopted by the ASTM are based on the "International System of Units" (designated SI for Systeme International d'Unites), fixed by the International Committee for Weights and Measures; this system is also known as the Giorgi or MKSA (meter-kilogram (mass)-second-ampere) system. This system has been adopted by the International Organization for Standardization in ISO Recommendation R 31.

The metric technical unit of force is the kilogram-force; this is the force which, when applied to a body having a mass of 1 kg, gives it an acceleration of 9.80665 m/sec/sec, the standard acceleration of free fall toward the earth's center for sea level at 45 deg latitude. The metric unit of force in SI units is the newton (N), which is defined as that force which, when applied to a body having a mass of 1 kg, gives it an acceleration of 1 m/sec/sec. These units must be distinguished from the (inconstant) local weight of a body having a mass of 1 kg; that is, the weight of a body is that force with which a body is attracted to the earth and is equal to the mass of a body multiplied by the acceleration due to gravity. However, because it is general practice to use "pound" rather than the technically correct term "pound-force," the term "kilogram" (or derived mass unit) has been used in this guide instead of "kilogram-force" in expressing the conversion factors for forces. The newton unit of force will find increasing use, and is essential in SI units.

QUANTITIES AND UNITS OF SPACE

Multiply	By	To obtain
LENGTH		
Mil.	25.4 (exactly)	Micron
Inches	25.4 (exactly)	Millimeters
	2.54 (exactly)*	Centimeters
Feet	30.48 (exactly)	Centimeters
	0.3048 (exactly)*	Meters
	0.0003048 (exactly)*	Kilometers
Yards	0.9144 (exactly)	Meters
Miles (statute)	1,609.344 (exactly)*	Meters
	1.609344 (exactly)	Kilometers
AREA		
Square inches	6.4516 (exactly)	Square centimeters
Square feet	929.03 (exactly)*	Square centimeters
	0.092903 (exactly)	Square meters
Square yards	0.836127	Square meters
Acres	0.40469*	Hectares
	4,046.9*	Square meters
	0.0040469*	Square kilometers
Square miles	2.58999	Square kilometers
VOLUME		
Cubic inches	16.3871	Cubic centimeters
Cubic feet	0.0283168	Cubic meters
Cubic yards	0.764555	Cubic meters
CAPACITY		
Fluid ounces (U.S.)	29.5737	Cubic centimeters
	29.5729	Milliliters
Liquid pints (U.S.)	0.473179	Cubic decimeters
	0.473166	Liters
Quarts (U.S.)	9,463.58	Cubic centimeters
	0.946358	Liters
Gallons (U.S.)	3,785.43*	Cubic centimeters
	3.78543	Cubic decimeters
	3.78533	Liters
	0.00378543*	Cubic meters
Gallons (U.K.)	4.54609	Cubic decimeters
	4.54596	Liters
Cubic feet	28.3160	Liters
Cubic yards	764.55*	Liters
Acre-feet	1,233.5*	Cubic meters
	1,233,500*	Liters

Table C–5 — Continued

QUANTITIES AND UNITS OF MECHANICS

Multiply	By	To Obtain
MASS		
Grains (1/7,000 lb)	64.79891 (exactly)	Milligrams
Troy ounces (480 grains)	31.1035	Grams
Ounces (avdp)	28.3495	Grams
Pounds (avdp)	0.45359237 (exactly)	Kilograms
Short tons (2,000 lb)	907.185	Kilograms
. . . . . .	0.907185	Metric tons
Long tons (2,240 lb)	1,016.05	Kilograms
FORCE/AREA		
Pounds per square inch	0.070307	Kilograms per sq. centimeter
. . . . .	0.689476	Newtons per sq. centimeter
Pounds per square foot	4.88243	Kilograms per sq. meter
. . . . .	47.8803	Newtons per sq. meter
MASS/VOLUME (DENSITY)		
Ounces per cubic inch	1.72999	Grams per cubic centimeter
Pounds per cubic foot	16.0185	Kilograms per cubic meter
. . . . .	0.0160185	Grams per cubic centimeter
Tons (long) per cubic yard . . .	1.32894	Grams per cubic centimeter
MASS CAPACITY		
Ounces per gallon (U.S.)	7.4893	Grams per liter
Ounces per gallon (U.K.)	6.2362	Grams per liter
Pounds per gallon (U.S.)	119.829	Grams per liter
Pounds per gallon (U.K.)	99.779	Grams per liter
BENDING MOMENT OR TORQUE		
Inch-pounds	0.011521	Meter-kilograms
. . . .	1.12985×10^{6}	Centimeter-dynes
Foot-pounds	0.138255	Meter-kilograms
. . . .	1.35582×10^{7}	Centimeter-dynes
Foot-pounds per inch	5.4431	Centimeter-kilograms per centimeter
Ounce-inches	72.008	Gram-centimeters
VELOCITY		
Feet per second	30.48 (exactly)	Centimeters per second
. . . .	0.3048 (exactly)*	Meters per second
Feet per year	0.965873×10^{-6}*	Centimeters per second
Miles per hour	1.609344 (exactly)	Kilometers per hour
. . . .	0.44704 (exactly)	Meters per second
ACCELERATION*		
Feet per second²	0.3048*	Meters per second²
FLOW		
Cubic feet per second (second-feet).	0.028317*	Cubic meters per second
Cubic feet per minute	0.4719	Liters per second
Gallons (U.S.) per minute . . .	0.06309	Liters per second
FORCE		
Pounds	0.453592*	Kilograms
. . . .	4.4482*	Newtons
. . . .	4.4482×10^{-5} * . *	Dynes

Table C–5 – Continued

QUANTITIES AND UNITS OF MECHANICS (Cont.)

Multiply	By	To Obtain
WORK AND ENERGY*		
British thermal units (Btu)	0.252*	Kilogram calories
	1,055.06	Joules
Btu per pound	2.326 (exactly)	Joules per gram
Foot-pounds	1.35582*	Joules
POWER		
Horsepower	745.700	Watts
Btu per hour	0.293071	Watts
Foot-pounds per second	1.35582	Watts
HEAT TRANSFER		
Btu in/hr ft² deg F (k, thermal conductivity)	1.442	Milliwatts/cm deg C
	0.1240	Kg cal/hr m deg C
Btu ft/hr ft² deg F	1.4880*	Kg cal m/hr m² deg C
Btu ft² deg F (C, thermal conductance)	0.568	Milliwatts/cm² deg C
	4.882	Kg cal/hr m² deg C
Deg F hr ft²/Btu (R, thermal resistance)	1.761	Deg C cm²/milliwatt
Btu/lb deg F (c, heat capacity)	4.1868	J/g deg C
Btu/lb deg F	1.000*	Cal/gram deg C
Ft²/hr (thermal diffusivity)	0.2581	Cm²/sec
	0.09290*	M²/hr
WATER VAPOR TRANSMISSION		
Grains/hr ft² (water vapor transmission)	16.7	Grams/24 hr m²
Perms (permeance)	0.659	Metric perms
Perm-inches (permeability)	1.67	Metric perm-centimeters

OTHER QUANTITIES AND UNITS

Multiply	By	To Obtain
Cubic feet per square foot per day (seepage)	304.8*	Liters per square meter per day
Pounds-seconds per square foot (viscosity)	4.8824*	Kilogram second per square meter
Square feet per second (viscosity)	0.02903* (exactly)	Square meters per second
Fahrenheit degrees (change)*	5/9 exactly	Celsius or Kelvin degrees (change)*
Volts per mil	0.03937	Kilovolts per millimeter
Lumens per square foot (foot-candles)	10.764	Lumens per square meter
Ohm-circular mils per foot	0.001662*	Ohm-square millimeters per meter
Millicuries per cubic foot	35.3147*	Millicuries per cubic meter
Milliamps per square foot	10.7639*	Milliamps per square meter
Gallons per square yard	4.527219*	Liters per square meter
Pounds per inch	0.17858*	Kilograms per centimeter

Table C-6. Determination of Wind Speed by Sea Conditions

Knots	Descrip-tive	Sea Conditions	Wind force (Beau-fort)	Probable wave height in ft.
0-1	Calm	Sea smooth and mirror-like.	0	-
1-3	Light air	Scale-like ripples without foam crests.	1	1/4
4-6	Light breeze	Small, short wavelets; crests have a glassy appearance and do not break.	2	1/2
7-10	Gentle breeze	Large wavelets; some crests begin to break; foam of glassy appearance. Occasional white foam crests.	3	2
11-16	Moderate breeze	Small waves, becoming longer; fairly frequent white foam crests.	4	4
17-21	Fresh breeze	Moderate waves, taking a more pronounced long form; many white foam crests; there may be some spray.	5	6
22-27	Strong breeze	Large waves begin to form; white foam crests are more extensive everywhere; there may be some spray.	6	10
28-33	Near gale	Sea heaps up and white foam from breaking waves begin to be blown in streaks along the direction of the wind; spindrift begins.	7	14
34-40	Gale	Moderately high waves of greater length; edges of crests break into spindrift; foam is blown in well-marked streaks along the direction of the wind.	8	18
41-47	Strong gale	High waves; dense streaks of foam along the direction of the wind; crests of waves begin to topple, tumble, and roll over; spray may reduce visibility.	9	23
48-55	Storm	Very high waves with long overhanging crests. The resulting foam in great patches is blown in dense white streaks along the direction of the wind. On the whole, the surface of the sea is white in appearance. The tumbling of the sea becomes heavy and shocklike. Visibility is reduced.	10	29
56-63	Violent storm	Exceptionally high waves that may obscure small and medium-sized ships. The sea is completely covered with long white patches of foam lying along the direction of the wind. Everywhere the edges of the wave crests are blown into froth. Visibility reduced.	11	37
64-71	Hurricane	The air is filled with foam and spray. Sea completely white with driving spray; visibility very much reduced.	12	45

from Weather Bureau Observing Handbook No. 1, Marine surface Observations, supersedes the Manual of Marine Meteorological Observations, Circular M, Twelfth Edition, March 1964.

CONVERSION CHART FOR PHI VALUES TO DIAMETERS IN MILLIMETERS

Table C-7 was reproduced from the *Journal of Sedimentary Petrology* with the permission of the author and publisher. It was taken from the Harry G. Page, "Phi-Millimeter Conversion Table," published in Volume 25, pp. 285-292, 1955. Includes that part of the table from -5.99 (about 63 mm) to +5.99 (about 0.016 mm) which provides a sufficient range for beach deposits. The complete table extends from about -6.65 (about 100 mm) to +10.00 (about 0.001 mm).

The first column of the table shows the absolute value of phi. If it is positive, the corresponding diameter value is shown in the second column. If phi is negative, the corresponding diamter is shown in the third column of the table. In converting diameter values in millimeters to their phi equivalents, the closest phi value to the given diameter may be selected. It is seldom necessary to express phi to more than two decimal places.

The conversion table is technically a table of negative logarithms to the base 2, from the defining equation of phi: $\phi = - \log_2 d$, where d is the diameter in millimeters.

Table C–7. Phi-Millimeter Conversion Table

ϕ	$(+\phi)$ mm.	$(-\phi)$ mm.	ϕ	$(+\phi)$ mm.	$(-\phi)$ mm.	ϕ	$(+\phi)$ mm.	$(-\phi)$ mm.
0.00	1.0000	1.0000	0.50	0.7071	1.4142	1.00	0.5000	2.0000
01	0.9931	0070	51	7022	4241	01	4965	0139
02	9862	0140	52	6974	4340	02	4931	0279
03	9794	0210	53	6926	4439	03	4897	0420
04	9718	0285	54	6877	4540	04	4863	0562
05	9659	0355	55	6830	4641	05	4841	0705
06	9593	0425	56	6783	4743	06	4796	0849
07	9526	0498	57	6736	4845	07	4763	0994
08	9461	0570	58	6690	4948	08	4730	1140
09	9395	0644	59	6643	5052	09	4697	1287
0.10	9330	0718	0.60	6598	5157	1.10	4665	1435
11	9266	0792	61	6552	5263	11	4633	1585
12	9202	0867	62	6507	5369	12	4601	1735
13	9138	0943	63	6462	5476	13	4569	1886
14	9075	1019	64	6417	5583	14	4538	2038
15	9013	1096	65	6373	5692	15	4506	2191
16	8950	1173	66	6329	5801	16	4475	2346
17	8890	1251	67	6285	5911	17	4444	2501
18	8827	1329	68	6242	6021	18	4414	2658
19	8766	1408	69	6199	6133	19	4383	2815
0.20	8705	1487	0.70	6156	6245	1.20	4353	2974
21	8645	1567	71	6113	6358	21	4323	3134
22	8586	1647	72	6071	6472	22	4293	3295
23	8526	1728	73	6029	6586	23	4263	3457
24	8468	1810	74	5987	6702	24	4234	3620
25	8409	1892	75	5946	6818	25	4204	3784
26	8351	1975	76	5905	6935	26	4175	3950
27	8293	2058	77	5864	7053	27	4147	4116
28	8236	2142	78	5824	7171	28	4118	4284
29	8179	2226	79	5783	7291	29	4090	4453
0.30	8123	2311	0.80	5743	7411	1.30	4061	4623
31	8066	2397	81	5704	7532	31	4033	4794
32	8011	2483	82	5664	7654	32	4005	4967
33	7955	2570	83	5625	7777	33	3978	5140
34	7900	2658	84	5586	7901	34	3950	5315
35	7846	2746	85	5548	8025	35	3923	5491
36	7792	2834	86	5510	8150	36	3896	5669
37	7738	2924	87	5471	8276	37	3869	5847
38	7684	3014	88	5434	8404	38	3842	6027
39	7631	3104	89	5396	8532	39	3816	6208
0.40	7579	3195	0.90	5359	8661	1.40	3789	6390
41	7526	3287	91	5322	8790	41	3763	6574
42	7474	3379	92	5285	8921	42	3729	6759
43	7423	3472	93	5249	9053	43	3711	6945
44	7371	3566	94	5212	9185	44	3686	7132
45	7321	3660	95	5176	9319	45	3660	7321
46	7270	3755	96	5141	9453	46	3635	7511
47	7220	3851	97	5105	9588	47	3610	7702
48	7170	3948	98	5070	9725	48	3585	7895
49	7120	4044	99	5035	9862	49	3560	8089

Table C–7 – Continued

ϕ	$(+\phi)$ mm.	$(-\phi)$ mm.	ϕ	$(+\phi)$ mm.	$(-\phi)$ mm.	ϕ	$(+\phi)$ mm.	$(-\phi)$ mm.
1.50	0.3536	2.8284	2.00	0.2500	4.0000	2.50	0.1768	5.6569
51	3511	8481	01	2483	0278	51	1756	6962
52	3487	8679	02	2466	0558	52	1743	7358
53	3463	8879	03	2449	0840	53	1731	7757
54	3439	9079	04	2432	1125	54	1719	8159
55	3415	9282	05	2415	1411	55	1708	8563
56	3392	9485	06	2398	1699	56	1696	8971
57	3368	9690	07	2382	1989	57	1684	9381
58	3345	9897	08	2365	2281	58	1672	9794
59	3322	3.0105	09	2349	2575	59	1661	6.0210
1.60	3299	0314	2.10	2333	2871	2.60	1649	0629
61	3276	0525	11	2316	3169	61	1638	1050
62	3253	0737	12	2300	3469	62	1627	1475
63	3231	0951	13	2285	3772	63	1615	1903
64	3209	1166	14	2269	4076	64	1604	2333
65	3186	1383	15	2253	4383	65	1593	2767
66	3164	1602	16	2238	4691	66	1582	3203
67	3143	1821	17	2222	5002	67	1571	3643
68	3121	2043	18	2207	5315	68	1560	4086
69	3099	2266	19	2192	5631	69	1550	4532
1.70	3078	2490	2.20	2176	5948	2.70	1539	4980
71	3057	2716	21	2161	6268	71	1528	5432
72	3035	2944	22	2146	6589	72	1518	5887
73	3015	3173	23	2132	6913	73	1507	6346
74	2994	3404	24	2117	7240	74	1497	6807
75	2973	3636	25	2102	7568	75	1487	7272
76	2952	3870	26	2088	7899	76	1476	7740
77	2932	4105	27	2073	8232	77	1466	8211
78	2912	4343	28	2059	8568	78	1456	8685
79	2892	4581	29	2045	8906	79	1446	9163
1.80	2872	4822	2.30	2031	9246	2.80	1436	9644
81	2852	5064	31	2017	9588	81	1426	7.0128
82	2832	5308	32	2003	9933	82	1416	0616
83	2813	5554	33	1989	5.0281	83	1406	1107
84	2793	5801	34	1975	0631	84	1397	1602
85	2774	6050	35	1961	0983	85	1387	2100
86	2755	6301	36	1948	1337	86	1377	2602
87	2736	6553	37	1934	1694	87	1368	3107
88	2717	6808	38	1921	2054	88	1358	3615
89	2698	7064	39	1908	2416	89	1350	4110
1.90	2679	7321	2.40	1895	2780	2.90	1340	4643
91	2661	7581	41	1882	3147	91	1330	5162
92	2643	7842	42	1869	3517	92	1321	5685
93	2624	8106	43	1856	3889	93	1312	6211
94	2606	8371	44	1843	4264	94	1303	6741
95	2588	8637	45	1830	4642	95	1294	7275
96	2570	8906	46	1817	5022	96	1285	7812
97	2553	9177	47	1805	5404	97	1276	8354
98	2535	9449	48	1792	5790	98	1267	8899
99	2517	9724	49	1780	6178	99	1259	9447

Table C–7 — Continued

φ	(+φ) mm.	(−φ) mm.	φ	(+φ) mm.	(−φ) mm.	φ	(+φ) mm.	(−φ) mm.
3.00	0.1250	8.0000	3.50	0.0884	11.314	4.00	0.0625	16.000
01	1241	0556	51	0878	392	01	0621	111
02	1233	1117	52	0872	472	02	0616	223
03	1224	1681	53	0866	551	03	0612	336
04	1216	2249	54	0860	632	04	0608	450
05	1207	2821	55	0854	713	05	0604	564
06	1199	3397	56	0848	794	06	0600	679
07	1191	3977	57	0842	876	07	0595	795
08	1183	4561	58	0836	959	08	0591	912
09	1174	5150	59	0830	12.042	09	0587	17.030
3.10	1166	5742	3.60	0825	126	4.10	0583	148
11	1158	6338	61	0819	210	11	0579	268
12	1150	6939	62	0813	295	12	0575	388
13	1142	7544	63	0808	381	13	0571	509
14	1134	8152	64	0802	467	14	0567	630
15	1127	8766	65	0797	553	15	0563	753
16	1119	9383	66	0791	641	16	0559	877
17	1111	9.0005	67	0786	729	17	0556	18.001
18	1103	0631	68	0780	817	18	0552	126
19	1096	1261	69	0775	906	19	0548	252
3.20	1088	1896	3.70	0769	996	4.20	0544	379
21	1081	2535	71	0764	13.086	21	0540	507
22	1073	3179	72	0759	178	22	0537	635
23	1066	3827	73	0754	269	23	0533	765
24	1058	4479	74	0748	361	24	0529	896
25	1051	5137	75	0743	454	25	0526	19.027
26	1044	5798	76	0738	548	26	0522	160
27	1037	6465	77	0733	642	27	0518	293
28	1029	7136	78	0728	737	28	0515	427
29	1022	7811	79	0723	833	29	0511	562
3.30	1015	8492	3.80	0718	929	4.30	0508	698
31	1008	9177	81	0713	14.026	31	0504	835
32	1001	9866	82	0708	123	32	0501	973
33	0994	10.0561	83	0703	221	33	0497	20.112
34	0988	1261	84	0698	320	34	0494	252
35	0981	1965	85	0693	420	35	0490	393
36	0974	2674	86	0689	520	36	0487	535
37	0967	3388	87	0684	621	37	0484	678
38	0960	4107	88	0679	723	38	0480	821
39	0954	4831	89	0675	825	39	0477	966
3.40	0947	5561	3.90	0670	929	4.40	0474	21.112
41	0941	6295	91	0665	15.032	41	0470	259
42	0934	7034	92	0661	137	42	0467	407
43	0928	7779	93	0656	242	43	0464	556
44	0921	8528	94	0652	348	44	0461	706
45	0915	9283	95	0647	455	45	0458	857
46	0909	11.0043	96	0643	562	46	0454	22.009
47	0902	0809	97	0638	671	47	0451	162
48	0896	1579	98	0634	780	48	0448	316
49	0890	2356	99	0629	889	49	0445	471

Table C–7 — Continued

φ	(+φ) mm.	(−φ) mm.	φ	(+φ) mm.	(−φ) mm.	φ	(+φ) mm.	(−φ) mm.
4.50	0.0442	22.627	5.00	0.0313	32.000	5.50	0.0221	45.255
51	0439	785	01	0310	223	51	0219	570
52	0436	943	02	0308	447	52	0218	886
53	0433	23.103	03	0306	672	53	0216	46.206
54	0430	264	04	0304	900	54	0215	527
55	0427	425	05	0302	33.128	55	0213	851
56	0424	588	06	0300	359	56	0212	47.177
57	0421	752	07	0298	591	57	0211	505
58	0418	918	08	0296	825	58	0209	835
59	0415	24.084	09	0294	34.060	59	0208	48.168
4.60	0412	251	5.10	0292	297	5.60	0206	503
61	0409	420	11	0290	535	61	0205	840
62	0407	590	12	0288	776	62	0203	49.180
63	0404	761	13	0286	35.017	63	0202	522
64	0401	933	14	0284	261	64	0201	867
65	0398	25.107	15	0282	506	65	0199	50.213
66	0396	281	16	0280	753	66	0198	563
67	0393	457	17	0278	36.002	67	0196	914
68	0390	634	18	0276	252	68	0195	51.268
69	0387	813	19	0274	504	69	0194	625
4.70	0385	992	5.20	0272	758	5.70	0192	984
71	0382	26.173	21	0270	37.014	71	0191	52.346
72	0379	355	22	0268	271	72	0190	710
73	0377	538	23	0266	531	73	0188	53.076
74	0374	723	24	0265	792	74	0187	446
75	0372	909	25	0263	38.055	75	0186	817
76	0369	27.096	26	0261	319	76	0185	54.192
77	0367	284	27	0259	586	77	0183	569
78	0364	474	28	0257	854	78	0182	948
79	0361	665	29	0256	39.124	79	0181	55.330
4.80	0359	858	5.30	0254	397	5.80	0179	715
81	0356	28.051	31	0252	671	81	0178	56.103
82	0354	246	32	0250	947	82	0177	493
83	0352	443	33	0249	40.224	83	0176	886
84	0349	641	34	0247	504	84	0175	57.282
85	0347	840	35	0245	786	85	0173	680
86	0344	29.041	36	0243	41.070	86	0172	58.081
87	0342	243	37	0242	355	87	0171	485
88	0340	446	38	0240	643	88	0170	892
89	0337	651	39	0238	933	89	0169	59.302
4.90	0335	857	5.40	0237	42.224	5.90	0167	714
91	0333	30.065	41	0235	518	91	0166	60.129
92	0330	274	42	0234	814	92	0165	548
93	0328	484	43	0232	43.111	93	0164	969
94	0326	696	44	0230	411	94	0163	61.393
95	0324	910	45	0229	713	95	0162	820
96	0321	31.125	46	0227	44.017	96	0161	62.250
97	0319	341	47	0226	426	97	0160	683
98	0317	559	48	0224	632	98	0158	63.119
99	0315	779	49	0223	942	99	0157	558

Table C-8. Values of Slope Angle θ and Cot θ for Various Slopes

Slope Angle θ	Cot θ (X/Y)	Slope (Y on X)
45°00'	1.0	1 on 1.0
42°16'	1.1	1 on 1.1
39°48'	1.2	1 on 1.2
38°40'	1.25	1 on 1.25
37°34'	1.3	1 on 1.3
35°32'	1.4	1 on 1.4
33°41'	1.5	1 on 1.5
32°00'	1.6	1 on 1.6
29°45'	1.75	1 on 1.75
26°34'	2.0	1 on 2.0
23°58'	2.25	1 on 2.25
21°48'	2.5	1 on 2.5
19°59'	2.75	1 on 2.75
18°26'	3.0	1 on 3.0
17°06'	3.25	1 on 3.25
15°57'	3.5	1 on 3.5
14°56'	3.75	1 on 3.75
14°02'	4.0	1 on 4.0
13°14'	4.25	1 on 4.25
12°32'	4.5	1 on 4.5
11°53'	4.75	1 on 4.75
11°19'	5.0	1 on 5.0
10°18'	5.5	1 on 5.5
9°28'	6.0	1 on 6.0
8°49'	6.5	1 on 6.5
8°08'	7.0	1 on 7.0
7°36'	7.5	1 on 7.5
7°08'	8.0	1 on 8.0
6°43'	8.5	1 on 8.5
6°20'	9.0	1 on 9.0
6°01'	9.5	1 on 9.5
5°43'	10.0	1 on 10.0

APPENDIX D

SUBJECT

INDEX

APPENDIX D

SUBJECT INDEX

— — A — —

— — B — —

-- C --

— — D — —

— — M — —

— — N — —

— — O — —

W — Continued

Wave—Continued

diffraction (see Wave diffraction)
dispersive 2-27
effects 2-1
 general 2-1
 on beaches 1-10, 1-11, 4-37, 4-115
energy (see also Wave energy) 2-27
energy spectra 3-11, 3-53
Finite Amplitude Theory (see also Wave, Trochiodal
 Theory; Wave, Stokes Theory) ... 2-3, 2-6, 2-7, 2-36
First-Order Theory (see Wave, Airy Theory)
frequency 2-4
 angular 2-9, 3-33
frequency of occurrence 4-110, 8-35, 8-36
fully arisen sea 3-34
general introduction 2-1
generation 3-15
 pressure pulse 3-15
gravity waves (defined) 2-4, A-15
 internal A-18
 seas 1-7, 2-4, 3-34
 swell 1-7, 2-4, 3-34
group velocity 2-24, A-15
height (see Wave height)
hurricane 3-52, 3-57, 8-53
 probable maximum wave 3-61, 8-54
internal A-18
irrotational A-18
kinetic energy 2-27
length (see Wavelength)
Linear Theory (see Wave, Airy Theory)
mass transport 2-4, 2-38, 4-4, 4-42, A-21
Michell Theory (maximum deepwater steepness)
 ....2-39, 2-120
monochromatic A-23
nonbreaking
 forces on piles 7-64
 forces on walls 7-127
nonlinear deformation 4-28
Nonlinear Theory (see Wave, Finite Amplitude
 Theory)
Oscillatory Theory (see also Wave, Airy Theory)
 ....1-7, 2-4, A-25
period (see Wave period)
periodic.................................. 2-4
phase velocity (see also Wave, celerity)........ 2-7
potential energy 2-27, A-27
prediction (see Wave prediction)
pressure (see Wave pressure)
pressure pulse wave generation 3-15
profile 2-8, 2-42, 2-114, 2-121
 group 2-26
progressive 2-4, A-27
refraction (see Wave refraction)
resonant 2-115
seas 1-7, 2-4
 fully arisen sea 3-34
shallow water (defined)................ 2-9, A-32
shoaling effects 2-121, 4-28
significant 3-2, 7-1, A-33
simple 2-4

Wave—Continued

Sinusoidal Theory (see also Wave, Airy Theory) ... 2-4
Small-Amplitude Theory (see also Wave,
 Airy Theory) 2-2, 2-6
Solitary Theory 2-2, 2-3, 2-59, A-34
standing (see also Clapotis; Seiche) 2-4, 2-113,
 ....3-78, A-36
stationary (see also Standing wave) A-36
statistics 3-5, 3-9, 4-4, 4-35, 4-36, 8-47
steepness 2-39, 2-120, 2-121
 maximum (for progressive) 2-39, 2-120
Stokes Theory 2-2, 2-3, 2-36
 profile 2-37, 2-41, 2-42
Stream Function Theory 2-62
swell 1-7, 2-4, 3-34, 3-42, 3-45
theories (see Wave Theories)
transitional 2-9, A-40
translatory 2-4, A-43
Trochoidal Theory 2-3, A-40
tsunami 1-7, 3-69, 3-70, 3-71, A-32, A-40
typical 3-2, 3-3
variability 3-5, A-41
velocity (see Wave, celerity)
wind generated 1-5, 2-1, 2-2, 3-15, 3-33, 4-29
Wave attack 1-5, 1-11, 4-29, 4-37, 4-70, 4-71, 4-115
 5-21
 protection from 5-49
Wave climate (see also Wave conditions) . 4-27, 4-28, 4-38
 effect on beaches 4-27
 nearshore 4-29
 offshore 4-28
 prediction (see also Wave prediction) 4-35, 4-36
Wave conditions (see also Wave climate) 4-28
Wave crest height (above SWL) 7-71
Wave data (recording) 3-2, 3-3, 7-2
Wave decay 1-7, 3-17, 3-42, 3-52, A-10
 deep water 3-42
 restricted areas 3-52
Wave diffraction 2-79, 8-93, A-10
 calculations, single breakwater 2-81, 8-93
 calculations, small gaps 2-98
Wave direction C-35
Wave energy 2-27, 3-11
 average 2-27, 3-5
 distribution by frequency (surf waves) 2-5
 kinetic.................................2-27
 longshore component 4-89, 8-107, 8-111, 8-122
 potential 2-27
 related to wavelength and wave height C-34
 spectra 3-11, 3-53
 total (oscillatory wave) 2-27
 total (solitary)...........................2-60
 transmission 2-27, 2-33, 2-66
Wave field decay 3-17
Wave forces 7-63
 effect of face slope 7-164
 incident wave angle 7-164
 Miche-Rundgren Theory 7-127, 7-128
 Minikin Theory 7-146
 nonvertical walls 7-164

D-13

— — Z — —